The Good Pets Guide

Alison Prince has found animals fascinating for as long as she can remember. She runs a small-holding in Suffolk and has cats, dogs, poultry and farm animals of every kind. She has worked as a writer and illustrator for many years and says her writing "grew up" with her three children, from a BBC "Watch With Mother" series, a "Jackanory" programme and several books for juniors, to her present output of novels for young adults and books about animals. She finds time in her busy life to fit in some teaching but prefers to keep clear of the day-to-day details of school routine. Her animals involve a lot of hard work but she says she would be lost without them. "Humans are much more difficult to deal with," she says cheerfully. "At least you know where you are with animals!"

THE GOOD PETS GUIDE

Written and illustrated by Alison Prince

An Armada Original

Among the many people who have been so helpful in the compiling of this book, my special thanks are due to Gill Standring of the London Zoo, to Pedigree Petfoods, to Michael Bignold of the Wildlife Rescue Service and to Phillip Glasier of the Falconry Centre. Countless others have been amazingly patient while I have taken up their time and gazed at their livestock, and I am grateful to all of them. For their unfailing helpfulness, I would like to add a particular word of thanks to the five splendid vets of Stowmarket, Suffolk.

The Good Pets Guide was first published
in Armada in 1981 by Fontana Paperbacks,
14 St James's Place, London SW1A 1PS

Printed in Great Britain by The Anchor Press Ltd,
Tiptree, Colchester, Essex CO5 0HD

Foreword

To add any living thing to one's household is a drastic step. A human baby makes the most dramatic impact of all, changing everyone's lives forever, but even a stick insect packs quite an emotional punch. Once it is there it has to be cared for. This is true of any creature kept as a pet, and no half-measures will do. Neglect your tadpoles and you will be faced with a stinky jamjar of rotting corpses. So think first. A fluffy puppy soon becomes a big, strong dog. Unless you can manage to train him and exercise him, don't take him on in the first place.

Animals are expensive, messy, demanding and fecund. They involve you in all sorts of work and trouble and they reproduce themselves with happy abandon. They can't be ignored, but for thousands of people they are a deeply necessary part of life. Animals as pets are there to be loved, controlled, cared for, watched with immense enjoyment, laughed over and wept for. But first make sure that you really mean business.

This book aims to help the would-be pet owner to find a suitable living creature – whether animal, bird, insect, fish, reptile, or even green pond water – which can be kept in an ordinary house and which will give pleasure and interest to the keeper.

Nobody can be truly happy with a pet unless the pet itself is happy, which means that it must be healthy and active, living in the best possible environment and eating the right food. In order to provide these things the owner must know in advance what his pet will need.

In an effort to provide concise, easily available information, this book is arranged in alphabetical order, each entry consisting of notes on the nature and needs of the species, including any special requirements as to the environment it needs to live in (temperature, humidity, etc.), its food, habits and reproductive ability. To avoid tedious repetition, a system of cross-references provides basic information on a group of creatures, with individual notes given under each separate heading. For example, house martins are described under their own heading, but the reader is also advised to see *Wild Birds*.

A star rating for each entry grades the "good pet" potential of each species from one star to five, with a cross for the utterly unsuitable. Details of the system are given on the next page.

Notes on the routine veterinary care of each species are given with particular reference to any necessary preventive injections which are needed, but it must be stressed that no book can be a substitute for expert advice. Any animal owner who is worried about his or her pet's condition must, of course, consult a vet without delay.

The Legal Bit

The laws about animals become more and more complicated as new bits are added to existing legislation in an effort to protect living creatures. A multitude of parliamentary Acts exists, but the average pet-owner is not affected by most of them. Dangerous wild animals such as bears, big cats and poisonous snakes are controlled by the Dangerous Wild Animals Act, 1976, which prohibits any private individual from keeping such animals without a licence. Endangered species are gradually coming under legal protection through a series of individual Bills which make it illegal to sell live or dead specimens without a licence. The otter, the greater horseshoe bat and the mouse ear bat, the sand lizard, the smooth snake and the natterjack toad are among the most obvious of protected species, and it is illegal to take any wild bird's eggs. Wild birds themselves may only be held captive if injured, and must be freed when fit to return to the wild.

You are very unlikely to be breaking the law if you buy livestock from a reputable supplier, who will be well aware of all legal requirements. The wild creatures of garden and stream will not land you in trouble with the law – but we must all respect the greater law of conscience and decent humanity. No animal should be taken from its natural life just as a plaything. Unless you can provide it with an acceptable environment, leave it where it is.

Star Ratings

***** Top marks – a completely satisfactory, companionable, interesting pet.

**** A very good pet with some slight limitation such as high purchase cost or specialised accommodation.

*** An interesting pet – often top of the range of less easy creatures to keep.

** A rewarding creature to study – many in this group are free-living, "voluntary" pets.

* Limited possibilities. Includes some of the less responsive creatures which may still be very intriguing.

X Don't keep these. Too wild, too dangerous, protected by law or plain unpettable.

ADDERS (Vipers) (see *Snakes*) X

Alligators X

Although looking like attractive little lizards when young, alligators grow into large, dangerous animals. Zoos already have plenty of them so it is best not to consider them as pets.

AMPHIBIANS

The name means "both-living" – a creature which lives in both air and water. Such animals actually *need* both elements, so if you want to keep newts or frogs or terrapins you must provide them with a proper vivarium. (see *Vivaria*)

Most amphibians start their lives in water and then become able to live on land. A frog is typical, living under water in his egg and tadpole stages then developing lungs as he evolves into his final shape. All amphibians continue to need water throughout their lives although mere dampness will do for quite long periods. They are cold-blooded (see *Reptiles*) and usually hibernate during winter. Their skins are very sensitive as they play an active part in enabling the animal to breathe, so never handle amphibians more than necessary. If you carry a newt about in your pocket it will die. Treat all amphibians very gently and carefully, and they can become rewarding pets.

Ants (see *Live Food*) **

There are two main kinds, black and red, though nearly thirty species exist in Britain. Red ants are carnivorous, eating small flies and aphids, and live in small colonies under flat stones. They bite humans sometimes. Black ants are milder natured, live in large underground colonies, are highly developed socially and "farm" aphids rather than simply eat them.

Ants are fascinating to study if kept in a formicarium where they can be seen. A glass tank or 5 litre (1 gallon) jar can house a fairly natural colony very easily. Find a black ants' nest in the garden and dig it up carefully with a trowel. In your glass jar put some earth from the nest, about fifty pupae (whitish-coloured), about 100 worker ants (small ones without wings, rushing about carrying pupae) and one wingless queen who will be about 1cm (½″) long. (Queen ants with wings are young and unfertilised but wingless queens lay eggs.) Quite soon the ants will start to organise chambers and tunnels in their new environment. Give them a sprig of rose or other plant infested with greenfly (aphids) and you can watch them "milk" these aphids of the juice which they secrete derived from the sap of the plant they live on. Ants appreciate a dab of honey on a spoon and need a little water – damp leaves will do. They will also eat dead insects. If you get tired of your pets, take the lid off the jar, put it on its side by the original nest, and they will all go home.

A more interesting though less natural observation nest is Lubbock's Formicarium, a glass/ants/glass sandwich kept on a moat-surrounded island to prevent escape. This probably results in neurotic ants but is fascinating to the scientifically minded. Construction is simple. Take a sheet of glass about 15 × 20cm (6 × 8″) and glue strips of balsa wood about 5mm (⅛″) thick all round the edges, leaving a 1cm (½″) gap in the middle of one side as an entrance hole. Within the balsa wood border, fill the glass with sieved, dampened earth. Now put a second sheet of glass (same size) on top to make your sandwich. Bind the edges all round except the entrance hole, with passe-partout (strong gummed paper used for mounting pictures). Glue a sheet of cardboard under the bottom glass and glue another sheet of cardboard to a third sheet of glass to act as a lid, for ants like to live in the dark.

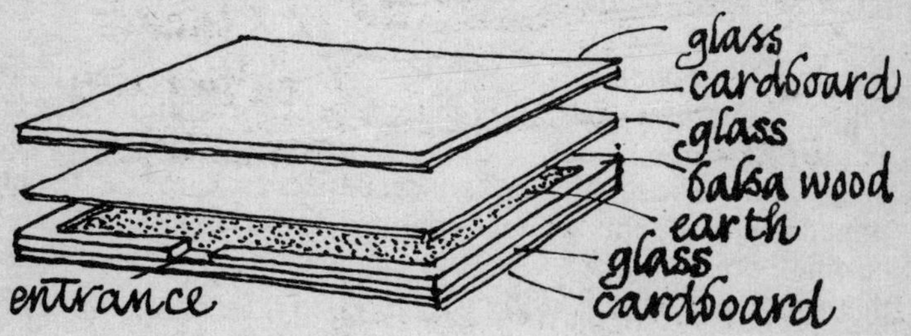

Now for the island. Find a baking tin and a slab of wood about 2–3cm (1″) thick, smaller than the tin but bigger than the formicarium. This will be your island. To stop it floating about, knock in some nails to keep it away from the sides of the tin. Fill the tin with water as a moat.

Now for the ants. How do you get ants into a formicarium? Simple. When you dig up your wild colony, put your selected workers, queen and pupae with some earth from their nest on a sheet of newspaper. Beside them, put your formicarium with its lid on. Desperate to get back into the dark, the ants will dive in through the entrance hole, carrying the pupae. When they are all in, put your formicarium on its island, standing it on 1cm (½″) slabs of cork so that the ants can run underneath it – and don't forget a cork "stepping stone" to the entrance. Leave the lid on for some hours until the ants have settled, and only take it off for short periods to look at them, for they will panic when light hits them. Remember to supply an aphid-laden sprig, and keep the moat full, or the ants will walk out.

This formicarium is cheap to make, free to run, quiet, interesting and uncanny in its resemblance to people in a modern city.

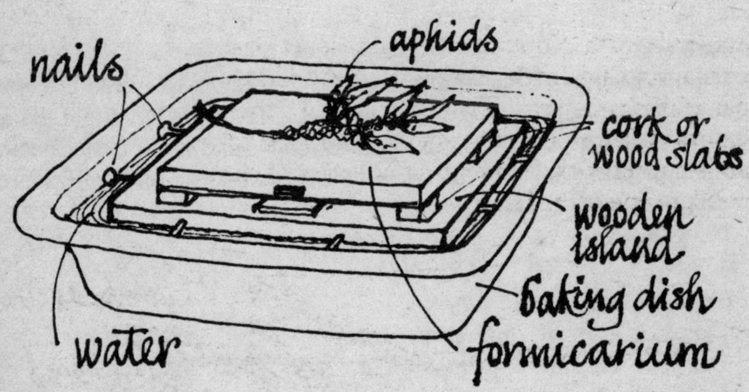

APHIDS (Greenfly etc.) (see *Ants, Live Food*)

Axolotls (see *Salamanders*) ***

A real weirdie. The axolotl is an overgrown salamander tadpole, 25cm (10″) long and still with gills, living in Mexico in the wild, or in a vivarium. It has a big head, four legs and a tail, and a fin down back and tail. Although technically immature, it breeds freely, laying eggs in spring which hatch in 2–3 weeks. Take the babes away or they may be eaten by their parents. Babies eat daphnia (water fleas) and adults like earthworms. (see *Live Food)*

The really curious thing about the axolotl is that in Mexico a long hot period can cause the rivers to dry up, baking the mud where axolotls live to a brick-like hardness. Anything else would die, but the axolotl simply loses its tadpole gills and develops into a mature salamander, independent of water. Ingenious!

Badgers X

Mainly nocturnal, these are wonderful animals to observe if you are lucky enough, but they are not pets. They are being gassed as carriers of bovine TB in the West Country and in time the species will no doubt be threatened with extinction, so badger families should be left to manage their affairs without any interference. They may only be kept under licence.

Bantams (see *Eggs, Poultry*) *****

These little birds, miniature versions of full-sized poultry, have much to be said for them. They are hardy and self-sufficient, coping much better with bad weather than their larger cousins, good breeders and good finders of food, even though it may be at the expense of your garden. They eat less than the bigger breeds and so are not expensive to keep.

The game breeds, slimly built, with wonderfully decorative cockerels, are apt to live quite independently, finding their own food and roosting in trees. I have seen them fly down from high branches on a winter morning so encrusted with ice that they hit the ground like footballs and yet come to no harm. At the other extreme, a very domesticated breed like the silkie, which looks like a powder puff, much prefers to be warm and dry. Silkie hens, incidentally, are marvellously motherly broodies, absolutely trustworthy with the eggs of other species.

There is quite a brisk trade in bantam hens each spring among gamekeepers who want a small, motherly bird to hatch out pheasant eggs. Full-sized hens are too big and clumsy.

Bantams, for some reason, tend to be pretty. Random cross-breeding results in attractive little birds in a vast variety of colours, though in-breeding often results in black becoming a dominant colour. Bantam cockerels are pugnacious little chaps and will often fight relentlessly, sometimes to the death. When this happens it is a proof of too many cockerels, so rescue the loser and give him to someone else or put him in the pot. Always keep the winner, for he is the stronger bird.

Bantam eggs are smaller than those of big hens but contain surprisingly large yolks. Since bantam cockerels are quick off the mark, cross-breeding with bigger poultry is common, and these half-breeds are very hardy and lay quite decent-sized eggs.

Even more than full-sized poultry, bantams are a much-fancied species for showing (see *Chickens*) and many of the most successful bantam fanciers are children.

See *Poultry* for details of how to keep bantams.

Barnacles (see *Sea Creatures*) **

Baby barnacles are free-swimming, only anchoring themselves to boatbottoms etc. when older. Don't try to scrape them off – chip out a piece of barnacle-encrusted rock if you want these for a marine tank. Good filter feeders, they will be very hungry in a clean tank.

Bats **

These are insect-eating mammals whose wings are really webs of skin stretched over very elongated fingers. They are nocturnal, detesting daylight, but have been kept successfully in captivity provided they have dark sleeping quarters for daytime use, with suitable rough ledges where they can hang upside down. Newly captured bats bite furiously but seldom inflict severe injury as they are no bigger than mice. The **pipistrelle,** or **common bat,** measures only 4cm ($1\frac{1}{2}$″) from head to tail. The **horseshoe bats,** known by the horseshoe-shaped lump on their noses, are a protected species and cannot be kept in captivity, but most other bats can be tamed and will eat raw meat, liver, hard-boiled egg and cream cheese as well as mealworms, cricket hoppers and other live food. In nature their food is always caught while flying. The young are born blind and sparsely furred, but well able to cling to their mothers, attaching themselves firmly to a teat. They begin to fly at about a fortnight old but are not fully independent for about two months. Bats hibernate in nature but wake on warm days to look for insects. In captivity they are best kept warm, well fed and awake.

Bees (also see *Bumble Bees*) ****

A single bee has to be regarded as a cell which belongs to the corporate body; the colony, with its collective mood or "hive mind" is the complete animal.

A colony of bees consists of a queen, who is the only egg-laying female, and her sterile female maids-of-all-work who nurse the infant grubs, collect nectar and pollen, keep the hive temperature right by fanning at the entrance with their wings and, of course, make honey. There are also some drones, an idle band of males who are there in case a mate may be needed for a young queen. When breeding ends in the autumn the workers chuck out the loafing lads to die in the cold.

A "swarm" of bees happens when the colony gets too big for the hive. Overcrowding triggers the bees' instinct to rear a new queen, and the canny beekeeper who inspects his hive every week will notice the big queen cells being built and will divide his colony into two hives, transferring some workers with a queen cell and destroying all unwanted queen cells to stop a population explosion. Unattended to, the old queen will sweep out in dudgeon with her retainers, leaving her usurping daughter in residence with the rest of the workers.

Capturing a swarm is a time-honoured way to set up as a beekeeper, but it requires nerve to knock the loudly buzzing football of bees off the branch, roof or hedge where they have settled. Even if you can bundle the swarm in a sheet and cart it home, you will have to persuade the bees to enter your prepared hive (assuming you have one). This "hiving" involves finding the queen among 30,000–40,000 bees and persuading her to go into the hive, whcn all the others will follow. It is easier to start off with a stock of bees already living in a hive, the whole thing bought from a reputable beekeeper. Contact the British Beekeeping Association (see *Useful Addresses*) who will put you in touch with your local Association. Beekeepers always seem to be amazingly friendly and helpful, and their experienced support is vital to a beginner, for there is a vast amount to learn. They will show you the various types of hive, teach you what your bees are doing, and sell you much cheaper equipment than you could buy direct from a supplier.

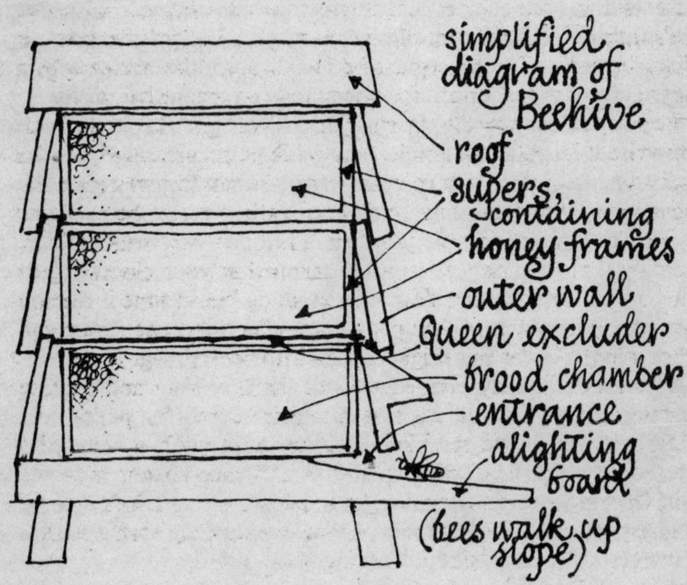

Bees themselves are not expensive but hives and equipment can be costly. Honey production in a good year can make a profit, but bees and their keepers are at the mercy of the weather. There is not much nectar about in a cold, wet summer when plants bloom reluctantly – and not much honey.

The basic hive, of whatever type, is used for breeding, as worker bees only live for three weeks at the height of the frantic summer when they slave away non-stop. In a wild colony, honey and "brood" cells would be all mixed together, but the beekeeper separates them by putting a queenproof mesh between the main hive and the upstairs "super", so his frames of honey are not muddled up with developing larvae. Several supers can be put on a hive in a good summer. Both supers and the main hive contain rows of frames side by side with just a "bee space" between them. These frames are the supports for sheets of wax ready stamped with the hexagonal shapes of bee-made cells, and on this "foundation" the bees build up their structure of cells which we call honeycomb.

In the autumn the beekeeper takes off the honey supers, extracts the honey in a centrifugal machine (borrow someone else's unless you keep a lot of hives) and puts it in jars for sale – a sticky business. The bees must be fed through the winter with a sugar syrup, for they made the honey to keep themselves alive.

Beekeeping is weirdly fascinating if you get the "bug" but otherwise it is quite impossible. You must be prepared to lever the sticky sections of your hive apart once a week to investigate the activities of your buzzing, crawling, rather cross bees. Some people do this clad in veil, gauntlets and boiler suit while others, claiming that bees only sting when trapped in your clothing, just wear a bathing costume. You *will* get stung but in time it doesn't hurt nearly as much. A few people are allergic to bee stings and either join the boiler suit brigade or give up beekeeping.

Bees are intimately connected with the flowering and fruiting processes of plants and trees as, in their search for pollen and nectar, they pollinate the blossoms they visit, with a wonderful effect on the "setting" of such things as runner beans and tree fruit. Orchard owners provide a large part of an apiarist's income by hiring a lot of hives for three or four weeks in the spring so that their trees are a-buzz with pollinating bees.

BIRDS (see *Birds of Prey, Oiled Birds, Orphans, Wild Birds* and individual headings for tame species such as *Budgerigars, Mynahs* etc.)

Birds of Prey (Kestrels *) X

Hawks and owls are by no means pets, and yet they are a fascinating study for anyone who is serious about them. Wild raptors, like all native British birds, are protected and neither they nor their eggs may be taken. It is now very difficult to get a licence to import them and the only way for a would-be falconer to obtain a bird is to buy a captive-bred one, which will, of course, be close-ringed.

If you are interested in these birds, see them in the hands of experienced owners and go to such places as The Falconry Centre (see *Useful Addresses*) run by Phillip Glasier, an eminent and very helpful falconer. Most people who keep hawks do so to fly them for game or "vermin" (rabbits etc.) in the traditional manner, but owls and kestrels are rather more confident than other birds of prey and lean a little more towards "pet" status.

Should you find an injured bird of prey, take it to an expert who knows what to do with it. The Wild Life Rescue Service in Norfolk (see *Useful Addresses*) is marvellous, but your vet may well know of someone living locally who keeps such birds and can cope with them. More harm than good is done by well-intentioned but ham-fisted help. Never give water to an injured raptor. The only safe food to offer is fresh raw meat, preferably a dead mouse or chick (hatcheries supply cull chicks in bulk to falconers), but a strip of raw beef will do. *Don't give it anything else.* The salts and minerals present in such things as bacon and tinned meats are totally foreign to a bird used only to its natural prey, and may kill it.

Different species vary enormously in appearance and temperament. The little **sparrowhawk** is hysterically shy and can't be kept in any kind of cage where it can see humans. The **kestrel**, by contrast, takes an intelligent interest in what goes on. Britain's commonest hawk, they are doing well on the mice and small birds which abound in the undisturbed grass verges and embankments of motorways. Kestrels have been breeding successfully in captivity for some time and for anyone seriously interested in birds of prey, this is the only species you are at all likely to be able to obtain. **Buzzards** – large, blunt-winged hawks – are a fairly lethargic lot. Moth-like in flight, they are solidly built and have a mewing call. They can reach 55cm (about 21″) in length.

The **peregrine falcon** is a great aristocrat, needing miles of open moorland to fly at game birds. Usually seen in mountainous places or on cliffs, these are about 40cm (15″) long with a sharp "kek, kek!" call.

Eagles are rare, and fiercely protected by the RSPB and every other right-thinking person. But do go and look at them in zoos, for they are truly magnificent.

Owls are subject to the same legal barriers as all other birds of prey and so are almost impossible to obtain. They are not used for hunting but, being more placid than the hawks, they often breed successfully in captivity. Although commonly called "night birds", they are in fact crepuscular – most active during the half-light of evening and dawn. They eat a variety of small birds and rodents, regurgitating the bones, feathers and fur in the form of "owl pellets".

Owls often choose to live near people, as they inhabit church towers or little-used barns, rearing their young on a flat surface such as a ledge. From the majestic **eagle owl** to that wickedly effective hunter, the **little owl,** this is a marvellous group of birds to study and admire, and it is not surprising that many old barns have an "owl window" high up in the roof to encourage these useful mouse-catchers.

To have owls – or any bird of prey – living freely nearby where you can watch them and enjoy their beauty rather than attempt to "keep" them, is a great joy. It is vaguely imagined that such birds can easily be rescued or hand-reared and then be set free – but the truth is far less palatable. To "hack back" a tame bird to the wild is heart-breakingly difficult, for the instinct to hunt and kill does not develop in a bird that has been fed dead chicks by its human keeper, and it is simply not able to fend for itself. The successful breeding in captivity of many species of owl may, however, mean that they may become a little more generally available as aviary birds.

Because these birds need expert knowledge, many falconers discourage any signs of interest from the public, knowing from experience what horrific mishandling birds of prey suffer from well-meaning amateurs. But if falconry is not to die out, a few dedicated newcomers to the art will be needed, and the really persistent will win through.

Bitterlings (see *Fish*) ***

About 8cm (3″) long, these are cold-water fish, the male mauve, pinkish or greenish with black-edged red fins, the female a quiet silvery-olive. They are interesting for their method of breeding, the female laying her eggs inside the open shell of a living pond mussel where they are fertilised by her mate. When the young hatch out they regard the mussel as their home, quickly swimming back into it if danger threatens. Since pond mussels need a layer of mud in which to live and move about, an outdoor pool is more likely to prove a successful environment than an aquarium.

Blennies (see *Sea Creatures*) ***

These are small rock-pool fish, easy to catch and keep in a marine tank, energetic and omnivorous, with a liking for chopped ragworm or mussel flesh. They will wriggle out of the water to bask on a projecting rock, which should be supplied. They have been known to move into a neighbouring tank because of dry-land locomotive skill.

BLOWFLIES (see *Live Food*)

BLUEBOTTLES (see *Live Food*)

Bluetits (see *Wild Birds*) **

These acrobatic little birds are great favourites with bird-lovers, for they are very pretty to watch and will readily bring up their families in a human-provided nest box as long as it is well sited and has an entrance hole of 3cm ($1\frac{1}{8}$″). A bigger hole will let bigger birds in. Bluetits and their even smaller cousins, the **coal tits** like a rich diet of nuts, seeds, bacon rinds and fat, and will readily come to a windowsill if it is temptingly festooned with nets of peanuts and dangling coconut halves. Like all British wild birds, they must not be kept in captivity.

Budgerigars (see *Cage Birds)* *****
Originally Australian, these are small members of the parrot family, living by nature in big flocks. They are strong fliers as they migrate annually from north to south Australia, avoiding the greatest summer heat. It is unkind to keep one alone, even more so if it is shut permanently in a small cage, for they need both exercise and companionship.

The best way to keep budgerigars is to build an aviary in the garden, providing a flight area and a sheltered sleeping place with perches and, in the spring and summer months, nest boxes for breeding birds. Budgerigars do not make nests in the wild, and need no nesting materials. Professionally constructed nest boxes have a shallow scoop in the bottom of the box where eggs are laid. The perch outside the entrance pop-hole is essential, for the hen bird incubating the eggs is fed by the cock perching outside the box. When the eggs hatch she in turn feeds the chicks with a regurgitated "milk" of half-digested seeds. Thus it is essential to see that an equal number of cocks and hens is kept, for it takes both birds to rear young. Sometimes a widowed hen will be adopted by a cock who will feed both her and his own wife, and orphaned young are often successfully reared by the one surviving parent.

Budgerigars will only breed if they are in prime condition, the male's "cere" (the fleshy patch above his beak) bright blue, the female's brown, and both birds clean and glossy. There is no reason why they should not achieve this, provided they have plenty of exercise and are well fed. As seed eaters, they are best fed on a mixture of canary seed and millet, preferably with some red rape, niger or sunflower seed added. Linseed is also good, and a bunch of seeding grasses hung in the cage will be enjoyed. Groundsel, chickweed, dandelion and salad vegetables provide valuable greenstuff and hard grit is essential for the gizzard to break up seeds. Cuttlefish "bone" and a mineral block should also be provided. Clean water must always be available, though budgerigars can manage without water for several days in their desert habitat. Aviary birds will sit in the rain to get their feathers wet so that they can preen properly and caged birds should be given a saucer of tepid water on a warm morning so that they can have a bath and be dry by evening.

Budgerigars should not be kept with smaller birds as they tend to bully. Among themselves they squabble about possession of the highest perches or nesting boxes, so these should be fixed high up in a level row to settle the argument. Eggs are laid from February to September, preceded by a lot of attentive behaviour, the cock offering the hen regurgitated food as though she were already incubating eggs. A solitary cock, poor chap, will do this to his image in a mirror. The hen lays 3–10 eggs, "sitting" from the first one laid. They hatch in eighteen days, so the first egg laid is the first hatched, and so on. The new-hatched chicks are bald and blind, opening their eyes in six days and being fully feathered in 28 days, at which point they start to leave the nest. Until their first moult they are striped across their foreheads. The first task is to learn to crack seeds with their beaks under their father's instruction. They can breed by four months but it is best to wait until the female is eleven months or more. Three clutches of eggs a year can be raised but this is exhausting for the parents and worries the owner, who can't find homes for the babies. Take the nest boxes out if breeding is not wanted, or remove the eggs as they are laid.

The basic colour for budgerigars is light green but there are lots of variations. They are cheerful, friendly little birds, living for

5–10 years and usually very healthy. Their claws need periodic clipping and they can be prone to developing tumours, which can often be removed safely if early veterinary advice is sought. They imitate human voices and other sounds, specially if patiently taught before they are six months old. Cock birds are better at this than hens.

Bullfinches (see *Wild Cage Birds, Wild Birds*) ***

Bullfinches produce apoplexy in gardeners by descending on the apple buds and pecking them to bits, but they are handsome birds with their bold, chunky heads and bright pink chests. Well-established as a captive strain, close-ringed specimens make tame pets, though they like seclusion and will not do in a "prison bars" cage. They need fruit, thistle heads, dock seed heads etc. and must have constant access to a bath. They will breed if conditions are right, making a nest of coarse, strawy materials.

Bumble Bees *

These are the big, furry wild bees, not to be confused with hive bees. If you can find a bumble bee's nest in the garden, watch it carefully, for it is a one-woman show, run by Mrs. Bumble alone, and it is just as interesting as a wild bird's nest.

A young, mated queen bee hibernates through the winter and, in spring, emerges, feeds and starts to build a nest, usually in a deserted mouse hole or other ready-made site. Her nest is tennis-ball-sized, made of dry leaves, grass, hair etc., and in it she lays up stores of nectar and of pollen, in which she lays her first eggs. There are about 8–16 of them and she broods them like a hen, keeping them warm. They hatch in five days and are then fed by their mother, who forages for nectar for them. It is five weeks before these first babies are full-grown bees able to help their mother with subsequent broods, and during this time many mother bees die through the exhaustion of trying to "cope". Bumble bee colonies seldom exceed 200, and all the eggs laid by the queen who was mated last autumn before hibernation will be females. Towards the end of the summer, this fertilisation weakens and she lays unfertilised eggs which hatch into males. These males eat very little and so nectar stores increase. This in turn triggers the production of queen cells, which are fed huge quantities of the stored nectar. These young queens stay in the nest, building up big stores of body fat to last them through hibernation. The old queen dies, as do all her workers, and the young males fly out to live freely, seeking young queens to mate with. Whether successful or not, they all die in the first frost, leaving only the fertilised queens to overwinter and continue the cycle.

Bumble bees are marvellous pollinators, being big and hairy and, because of an amazing ability to raise their own body temperature at will, are able to fly when it is too cold for hive bees. For this reason they are excellent things to have in the garden. They are subject to drunkenness if they get at the lime tree blossom and there is a "pirate" bee known as a "cuckoo bumble" which takes over other bees' nests. A bumble bee's sting is unbarbed, so she will not die if she stings you, but they are not aggressive and should be left to conduct their hard-working lives in peace.

Bushbabies ***

There are several kinds but the commonly sold one is the **Senegal bushbaby,** a small, furry mammal with big eyes, a long tail and natty little feet.

Being nocturnal and African, bushbabies like it warm and dark. They need a lot of space and cannot be house-trained, so they are tricky little things to keep. The ideal housing would be a straw-filled, centrally heated spare room, for the temperature must not fall below 18°C (65°F) and should not fluctuate much. Bushbabies are very active after dusk and need branches to leap about on. If let loose in the living room, dim the lights, shut the window, block up the chimney and remove everything breakable, for they can jump about 2 metres (6′).

Bushbabies eat insects and fruit, so keep a supply of mealworms (see *Live Food*) and provide an assortment of dates, sultanas, grapes, bananas, oranges, greenstuff and corncobs. They also like bread and milk, milk alone and occasional condensed milk, and must have a constant supply of clean water.

Breeding in captivity is not common. The gestation period is about four months and usually two young are born, well furred and with their eyes open. The mother carries her young about with her but for the first few weeks it is wise to house dad separately. Baby bushbabies start to take an interest in solid food at four weeks and eat like adults at eight weeks, though they continue to suckle for some time.

Bushbabies are expensive animals to buy and almost impossible to sex, so professional bushbaby breeders are rare.

Butterflies (also see *Silkmoths*) ***

It is surprisingly easy to rear butterflies and can even be profitable, since professional suppliers are interested in buying back surplus stock from breeders. (see *Useful Addresses*).

A vast range of species can be reared, but several things must be remembered. First, make sure that you have ample supplies of the plant or tree whose leaves your chosen caterpillars eat. Don't choose something which has to eat oak leaves if the nearest oak tree is three miles away. Caterpillars can die in as little as twenty minutes if they run out of food, so a ready supply is absolutely vital. Second, decide what you are going to do with the adult butterflies before you start hatching eggs. If you intend to release them into the wild, choose something which is not going to savage the neighbouring gardens. Several hundred **cabbage whites** would be enough to cause any gardener apoplexy. Think also about the natural vegetation in your district. If you are going to release stock to live naturally, there must be food available.

Ecology is never as simple as it seems. If a particular butterfly is rare in your district, its numbers must be built up gradually. To release thousands of them will simply attract the inevitable predators to an unusual feast, and they will probably gobble up not only the newly released stock but those already living in the wild as well. Drastic changes to the balance of nature are always to be avoided.

A supplier will send rearing instructions with the eggs you order, but in general, you can expect to hatch 10–20 eggs in a matchbox. The tight-fitting lid will preserve humidity and there is quite enough oxygen. Just open the box and breathe in it once a day, and all will be well. Most species hatch in a few days. The tiny caterpillars must not be touched with your fingers, for there is enough acid in human skin to kill them. Use a fine paintbrush to transfer the larvae to a box supplied with host plant leaves. Never give water – the leaves contain enough moisture. Make sure the host plant has not been sprayed with insecticide. Butterflies, after all, are insects. Wash any suspect leaves but shake them dry before use.

Caterpillars will change their skins on the second or third day, and four more times before they are big enough to pupate. Although butterflies do not generally hibernate, they have a biological "rest period" known as the diapause, when they go into a state of suspended animation. Some do this in the larval stage, some as pupae and some as adults, but it is always a mechanism controlled by daylight length, designed to ensure that the insect reaches the right stage of its life at the right time when food plants are available.

An easy species for the beginner could be the **small tortoiseshell** – the red, brown and purple butterfly which feeds on the nectar of buddleia blossoms or michaelmas daisies. This lays 40–120 eggs on the back of stinging nettle leaves. The caterpillars grow to 4cm (1½″) long, black in colour with white flecks. They pupate and emerge as adults ten days later.

Adult butterflies can be kept as breeding stock in fine mesh cages with plenty of the host plant on which to lay their eggs. The adults never eat leaves, and their normal diet of nectar can be substituted very well by a honey and water mix.

Moths are a much larger group than butterflies but their nocturnal habits make them less suitable for indoor culture. The glamorous **silkmoths** are a different thing altogether and notes on rearing them are included under a separate heading.

BUZZARDS (see *Birds of Prey*) X

Cage Birds (see individual headings) ****

It would be much pleasanter to describe these beautiful little creatures as "aviary birds", for it is thoroughly distasteful to see any living thing cooped up in a prison where it is unable to move freely and live anything approaching a normal life. If a cage must be used, it should be as big as possible and a solitary pet bird should be allowed out to fly freely every day – making sure, of course, that windows are shut.

As a general rule, birds should be kept in pairs, for the males will fight over the females if a lot are kept together, and there is a lot of squabbling about nest boxes and nesting materials. Most species need an "indoor" area in their aviary to give them shelter from bad weather, and they must all be provided with clean, fresh water for drinking and bathing. They must have grit and "cuttlebone" as well as a variety of seeds such as millet, millet sprays, hemp, canary seed, red rape seed, niger and seeding grass heads. Lettuce and hard-boiled egg are also liked, and live food such as moths, beetles and the ever-useful mealworm (see *Live Food*) should be given whenever possible.

Newly imported birds have come from a warmer climate and will need to get used to our chillier weather slowly. Start them off indoors in a thermostatically controlled room between 18–21°C (66–71°F), dropping the temperature gradually, a little each day. Many species are protected by law – including all native British birds. It may come as a surprise to see such species as the **bullfinch** and the **linnet** still sold as cage birds, but this can happen because for many years these birds have been captive bred. The young of such domestic strains may legally be sold and must be close-ringed as fledglings. Nearly all pet birds are seed-eaters and notes on individual variations are given under species headings. **Mynah birds** are fruit-eaters and need different treatment.

Some good cage or aviary dwellers are: **cardinals, twites, redpolls, silverbills, java sparrows, mannakins** and the many species of **finch. Canaries** remain deservedly popular and notes on these appear under their own heading.

CALVES (see *Cattle*) X or ****

Canaries (see *Cage Birds*) ****

So called because they came originally from the Canary Isles, these remain a favourite pet bird because of their tameness and the sweet song of the male (the female just chirps). The **roller** or **harz canary** is the best singer, always performing with its beak shut. A louder, open-beaked species is the **Belgian Malinois** or **waterslagers.** Rollers are usually variegated and the Malinois a clear yellow. "Colour canaries" are now being bred, but the red-factor birds have to be given special food to retain the red colour. Other variants – or perversions – have changed the bird's shape to give it crests and frills.

A pair of canaries will breed if given ample-sized cages where they can fly, laying their eggs in a nest of hay or grass, hair, moss or felt. They will need a good supply of this kind of material and, unusually, prefer an open nest box, which is pleasant for the interested owner. Crumbled biscuit and hard-boiled egg should be added to the normal diet when the birds are rearing young, and when the baby birds first come out of the nest it is best to feed only this soft crumble until they are big enough to manage seeds.

Canaries can rear three or four broods in a season and live for up to 25 years.

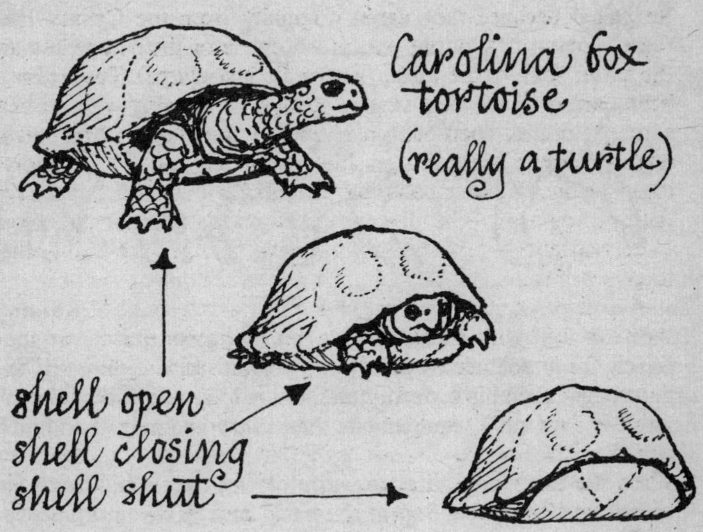

Carolina Box Tortoises (see *Terrapins, Tortoises*) ***

Confusingly, this is really a terrapin but one which has become mostly land-dwelling. It looks like a tortoise except that the underside of its shell is hinged, enabling it to shut up like a box.

Like most terrapins, this creature is carnivorous, eating insects, slugs and worms. It also eats fruit and mushrooms but no greenstuff, which makes it a gardener's friend, unlike the lettuce-chomping tortoise. The female has brown eyes, the male red ones. They are egg-laying reptiles, better adapted to cool conditions than the terrapins or the Mediterranean tortoise, and prefer to hibernate at the bottom of a garden pond, provided it is nice and muddy.

CATERPILLARS (see *Butterflies, Moths, Silkworms* etc. also see *Live Food)* * or **

Cats *****

To get on well with a cat, one has to recognise that he or she belongs to a solitary-dwelling, self-reliant race. He likes to organise his life in his own way and, in the long run, it is easier to fit in with your cat's requirements than to try and make him fit in with yours.

Cats like two meals a day, a warm place to sleep and freedom to come and go. A cat-flap in the back door saves an immense amount of time and trouble in putting the cat out or getting him in. It does, of course, permit muddy pawmarks all over the kitchen floor, but there is no solution to that problem short of stationing a commissionaire armed with a towel by the cat-flap.

The risk of free-living cats being run over is one which cat-keepers simply have to take. Pedigree breeders keep their stock in cages but domestic pet cats cannot be deprived of their liberty in this way. Some cats learn to understand the dangers of the road and some, with fatal results, do not. It is deeply distressing to have one's pet killed outside the house but it is a risk balanced by the great pleasure of a cat's companionship. Life cannot be completely safe for any of us.

This lack of controllability is something which many people find difficult to accept about cats – and yet, once it is completely admitted, it forms the basis of a marvellously unforced friendship between owner and pet. Cats which are free to lead their own lives have a spontaneous affection for their human companions which is immensely rewarding.

Cats, like humans, form most of their ideas during childhood, so be careful what impression of yourself you give to a newly arrived kitten. If it does not completely trust and like you, there is no hope of it ever being a truly affectionate cat. The biggest wrecker of relationships is, of course, house training. It is quite natural to a kitten to defecate in a remote corner of a room rather than venture out into an unknown, cold, perhaps wet, garden. It is natural also to hate the human who descends, shouting, to ram the kitten's nose in a pile of excrement and hurl it outside. Conversely, a human who will take the kitten outside within quarter of an hour after eating, put it on some soft earth and stay with it until it has obliged, is being very reassuring and deserves an adoring cat.

Kittens are ready to leave their mother at eight weeks old. They should be energetic and bright-eyed, and their mother should be calm and not too thin. Avoid the kittens of a haggard, suspicious, flea-ridden cat housed in some damp shed, for such a start is a fearful handicap. Both male and female kittens can be neutered if breeding is not wanted, but you need to know the sex of your kitten. If you can lay your index finger across the space between the two openings (anus and genital), then the kitten is a male. If not, it is female. With fluffy kittens this can be quite difficult to see, but if you can't tell, it's usually a female. Don't be tempted to keep an un-neutered Tom – he will make awful smells, stray away and get into fights, returning with nasty injuries. Have him neutered between four and six months.

Long-haired kittens look wonderful but need unending care. Daily brushing and combing is needed if their coats are not to turn into a matted mess, and hair balls will form in the stomach if long-haired cats are left to groom themselves.

Feed your newly arrived kitten at least three times a day with small feeds of the tinned or fresh food it has been used to. Always ask the breeder what your kitten has been eating, and never change its diet drastically or it may become ill. Have fresh, clean water available as well as milk, but remember cats won't touch anything stale or dusty. Tinned foods are a good basic diet but cats love raw meat. For kittens, either give a big lump to chew at or cut it finely – avoid mouth-sized pieces which can choke them.

A female ("queen") cat born in the spring will be a year old before she comes in season the following spring, but a summer-born kitten will also come in season the next spring, when she will be perhaps barely eight months old. It is difficult to restrain a "calling" female, for cats are extremely determined to mate and if prevented from doing so, will come into season again about fourteen days later. They will keep doing this until successfully mated. Pedigree cats in particular tend to yowl horribly and climb up the curtains if shut up. Because of this, some breeders use the hormone "pill" to control ovulation but this is not a total solution to the problem. Ask your vet by all means, but basically your cat either has kittens when she wants to, or you have her spayed and she doesn't have them at all.

The gestation period is usually 63 days, the litter size on average 3–5, often more. Cats seldom need help in kittening and the young are fully furred though their eyes do not open for about ten days. An older queen will often help her daughter kittening for the first time, ushering her into the prepared box and biting through the umbilical cord of the first kitten born, though they usually seem to let the daughter cope on her own after that.

A pregnant cat will hunt about for a good place to have her kittens, and if she is to use the box you provide, it is essential to put it in a place the cat likes. See it from her point of view and make sure to pick a warm place, free of draughts and, most important, private. In nature the mother cat must defend her kittens against predators and the pet cat feels just the same. She will like a tucked-away spot in the bottom of a cupboard or the corner of the kitchen where her kittens are not easily seen. She will want only one door giving access to them so that she knows where "enemies" are coming from. And if friends and neighbours flock in to handle and coo over the kittens, she will move them to a safer place. Some cats like to have their babies in a box placed on a high surface for good defence, but these will always move their families to floor level at about three weeks old when they may creep out of the box or basket.

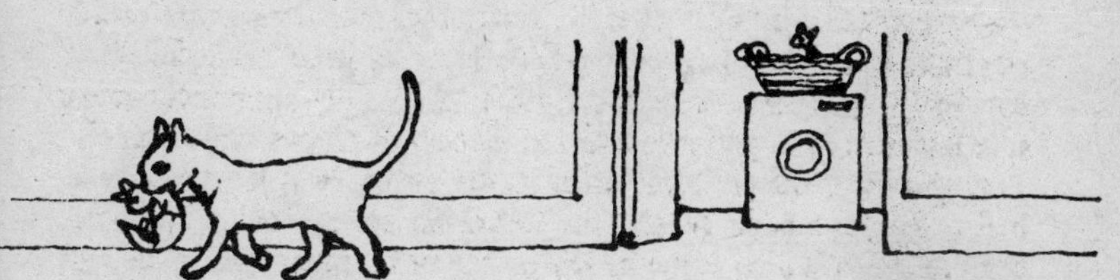

Clean newspaper is the best lining for a kittening box. It is warm and cannot catch little claws to pull limbs in the wrong direction. It also can – and must – be changed frequently, choosing a time when mum is out so that she does not see her kittens being handled. Fleas, if present, flock on to kittens because of their tender skins, and although these pests can be killed by powdering with a safe product (ask your vet), their eggs develop on bedding, not on cats. Clean bedding keeps cats flea-free – and this is true for adult cats as well.

Another pest is the presence of parasitic worms. These affect almost all cats, particularly those which hunt, as the tapeworm is carried by mice. Roundworms are passed from cat to cat, and both of these parasites cause thinness and loss of condition. A bad infestation can result in anaemia and nervous disorders as well as interfering seriously with the digestion, so all cats must be wormed regularly – at least every six months, oftener for hunting cats. Your vet will advise on which preparation to use. The dose is often based on body weight, and the easiest way to weigh a cat is to weigh yourself with and without cat on the bathroom scales. The difference is the cat's weight. To dose a cat, put your hand over the top of its head and open its mouth with one finger. With a quick poke, pop the tablet well in, over the ridge of the tongue. Hold the cat's mouth shut and stroke its throat until it has gulped. If it fights, wrap it firmly in a towel, tucking all paws in. For some cats, a helper is essential.

Feline infectious enteritis and cat flu are killer diseases preventable by vaccination – ask your vet. Cats have a wonderful ability to heal injuries but illness lays them low very rapidly, so seek advice at the first sign of sneezing, runny eyes, obvious droopiness or diarrhoea. The vast majority of cats are healthy and resilient and, given warm sleeping quarters and good food, will live for 12–16 years.

Cattle **(House Cows ****)** X

Cattle are not pets. They require specialist housing and handling and are subject to all sorts of regulations. This being said, a house cow kept by country-dwelling people who know what they are doing can be a most amiable and deeply satisfying companion. She is also a way of life, for everything has to be organised round her regular feeding and milking.

A house cow need not depend on grass as her staple diet but she must have enough open space for exercise. She needs a large tankful of clean water, a shelter against bad weather and plenty of straw as a bed. She can eat a variety of rolled cereals with some bran, sugar beet pulp nuts or proprietary pellets – every cowman has his own chosen mixture. She also needs hay and will benefit from some root crops such as swedes or mangels. From two years old onwards she should calve each year, the local artificial insemination service (and luck) achieving this. A **Jersey** gives the highest butterfat percentage, is hardy and gives the best value for feed. Cross-breed her with a beef breed for a chunky calf to sell or fatten for the deep freeze. **Aberdeen Angus** gives a small calf for a first calving but **Murray Grey** is equally good. A **Charolais cross** often gives excellent beef.

Do not take on a house cow unless someone in the family is experienced, or cow-keeping friends live within shout-for-help distance. Mistakes can very easily be fatal and good cows are worth several hundred pounds.

Jersey cow with beef-type calf

CAVIES (see *Guinea Pigs*) ****

Centipedes *

Fast-moving creatures with one pair of legs to each body segment, and long antennae. Carnivorous, eating soft-bodied insects, slugs etc. A good friend to the gardener but must be supplied with plentiful live food if kept in captivity.

Chameleons (see *Lizards*) **

One of the best-known reptiles, this is not, in fact, a very rewarding pet as it hates being handled and is rather tricky to keep. There are about fifty chameleon species but only two are at all suitable for a domestic vivarium – the **common chameleon** from North Africa or Spain, and the South African **pygmy chameleon.** This is 15cm (6″) long, usually a dull yellow-green colour which can change to a dull blue-green or an equally dull blue-mauve. This takes quite a time to happen, incidentally, and it's no good putting a chameleon on a draughts board and expecting it to go black and white. It can't. Neither can it swim, so don't have any open water in the vivarium (see *Vivaria*) or it will drown. It can't drink except by sipping dew from leaves (see *Dew Dispenser*) and it only eats live flying insects, no worms or grubs or anything which keeps still (see *Live Food*). It needs constant sunbaths, so must have an overhead light. It will breed in captivity but has a delayed implant mechanism, which means that it's anybody's guess how long after mating the live young will be born.

Despite all disadvantages, the huge boggle eyes are rather attractive – their independent suspension means they can look two ways at once – and a chameleon's feet have a "thumb" on the opposite side from the "fingers" so that it can grip things as we do. Its tongue ends in a sticky bulb, enabling it to snatch up flies too fast for the eye to follow.

Cheetahs X

These used to be thought of as classy pets for snooty girls, but their sale is now controlled by law. They are bred in captivity but getting a licence to keep any such animal is not easy.

Chickens (see *Poultry*) ****

There is an immensely wide variety of breeds available, some designed mainly as egg-layers and others, like the big white **Cobb,** intended as table birds. Chickens are also much fancied as show birds, however, and there is a great interest in breeding pedigree stock or in cross-breeding to produce attractive-looking hybrids. **Aracaunas**, for instance, are a South American breed with tufts on their heads, notable for laying pale blue eggs. Crossed with the French **Maran**, a dark brown egg-layer, the progeny lay delicate green eggs. The list of interesting breeds is endless.

The summer agricultural shows which take place in every county are a rich hunting-ground for the stock fancier, for there is always a "Fur and Feather" section with competitions for poultry, cavies and rabbits, and this is a great chance to meet breeders and find out more about the birds you fancy. There is no difference between keeping ordinary back-yard poultry and a specialist breed except that you must keep a pure-bred cockerel with his own hens and not let them mix with other kinds, which may involve a little more wire netting.

The modern hybrids are evolved for battery use, where continuous egg-laying is wanted and broodiness is a crime punishable by death. **Warren Studlers, Arbor Acres** etc. will lay very well but are useless to anyone intending to breed poultry. The **Rhode Island Red X Light Sussex** is a big brown hen which does everything – lays, broods and provides a good Sunday dinner. But decide first what you want and don't settle for anything else, for chickens are as different from each other as the various breeds of dogs are.

For details of how to keep chickens, see *Poultry*.

CHICKS (see *Eggs, Poultry*) ****

Chipmunks ***

Quick-moving, shy, squirrel-like animals, these are known by several names, most commonly the **Siberian chipmunk** but also the **Korean pygmy squirrel,** the **striped ground squirrel** and the **Japanese miniature squirrel.** In the wild they dig out burrows with special food stores stuffed with nuts, dried fruits, berries and seeds against the winter, which they spend eating and sleeping in their underground bedrooms, though they do not actually hibernate.

If thoroughly acclimatised to British weather, chipmunks can be kept in outdoor cages provided they have well-boarded, warm sleeping quarters at one end. Indoors, they are all right at normal

room temperatures though they may be very torpid during winter months. Being extremely agile and escape-prone, they must always be caged except for supervised exercise periods, but cages must be over 1 metre (about 4′) high, supplied with branches to climb on and nesting boxes 25cm (9″) each way, with access through a pop-hole halfway up one side. Provide a box for each animal, and supply straw or newspaper as bedding.

Chipmunks eat peanuts and other nuts, corn, millet, sunflower seeds etc., apples, carrots, greens and all kinds of dried fruit. They also like occasional mealworms (see *Live Food*). They must gnaw to prevent their teeth from overgrowing, so need fresh, green-barked wood such as tree prunings.

A pair of chipmunks live more happily than a solitary one, but take care in introducing a strange animal to an established one, for it may be attacked. They will only breed if the cage is large enough to be a complete habitat with hiding places. The gestation period is about thirty days, the mother looking after four or five young very well if she is left in complete privacy. Never look in the nest box, and house dad separately for about five weeks.

COCKATIELS (see *Cage Birds, Parrots*) ****

COCKATOOS (see *Cage Birds, Parrots*) **

Cockerels (see *Bantams, Chickens, Eggs, Poultry*) ****

Although apt to wake you at dawn, cockerels have their uses. They prevent quarrelling among the hens and tend to keep them together, and they always behave like gentlemen, never hogging any titbit they find for themselves but always calling their hens to it. What's more, a bold, glossy cockerel is a constant pleasure to look at.

Cockroaches **

These are ancient prehistoric insects, frowned upon by health authorities because they like to live in bakeries, kitchens and cellars – anywhere warm and damp. This is because they are of tropical or subtropical origin, only two of their six species being British, and these two both now very rare. The others have arrived via the holds of ships.

Cockroaches can be kept in glass jars with sawdust on the bottom, so damp that the glass is dewed with condensation. Keep over a light bulb or near a hot tank, as the minimum temperature must be 20°C (70°F). They will eat anything organic – bread, cake, sugar, dog biscuits, flour, potatoes or even newspaper – and breed freely in the right environment, the egg capsule being dropped rather casually in some corner. The nymphs are wingless and moult several times. Any casualties are tidily eaten up by the others.

COWS (see *Cattle*) X or ****

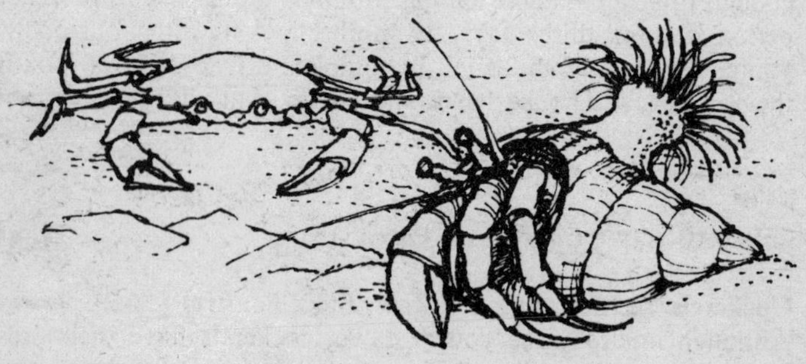

Crabs (see *Sea Creatures*) ***

Efficient, belligerent scavengers, these eat anything and everything. A small one is a good living dustbin in your marine tank but a small "crabs-only" tank is the best way to keep a number of them. Hermit crabs are very interesting as these, lacking a shell of their own, move into an empty shell of another species, often sharing it with a ragworm and a sea anemone, the ragworm inside and the anemone stuck on the roof. When the shell becomes a tight fit, the hermit crab has to move house and find a bigger shell. Despite the happy coexistence in early days, a large anemone will often eat a hermit crab, so it is best to keep big specimens apart.

Crested Anolis (see *Lizards*) ***

23cm (9″) long, the anolis is dark blue with a black spot on each side and a crest down the back and along its tail. Its toes are cross-ridged like tyre treads, enabling it to run up walls. Since it naturally lives near water, it must have a pool in the vivarium (see *Vivaria*) and a variety of live insects (see *Live Food*) to eat. It is all right at room temperature during warm months but will need heat in the winter. The anolis is a rather nervous creature and when upset will inflate its throat sac to a large balloon – rather like the effect of teasing dad, only more so. Caution – if you own one of these, don't let your friends poke things at it just for the fun of seeing it puff up. Once it has confidence in you it can become rewardingly tame, but not if it is constantly teased.

Crickets (see *Live Food*) **

An excellent live food for amphibia and most lizards, these are easy to breed. They need a temperature of 21–26°C (70–80°F) and can be kept in large glass jars or plastic boxes. Low temperatures may make them appear to be dead but they will revive when warmed up. Feed them on bread – or wholemeal flour for young ones – and pieces of apple to provide moisture. Keep big crickets separate from little ones as they eat each other.

Provide a dish of damp sand for egg-laying. New-hatched crickets emerge in about a fortnight and change skins several times. These small "hoppers" are good for carnivorous insects and young reptiles. Fishing shops supply crickets or they can be bought from specialist suppliers. (see *Useful Addresses*)

Crows (see *Wild Birds*) *

Unlike rooks, which are flock birds, crows live alone or in pairs. They are large, strong birds, eating mostly carrion but also eggs, insects and grubs and occasional small birds. The **raven** is a member of the crow family, famous for its presence at the Tower of London. Ravens, like all crows, are highly intelligent and make attractive pets in a voluntary sort of way, remembering that no bird may be taken from the wild. **Jackdaws** have long been popular as mischievous pets, well known for their habit of picking up shiny things. They are only half the size of the big raven, a mere 35cm (14″) long, and can be distinguished by the grey back to the head and neck. A naturalist called Konrad Lorenz is famous for his work with members of the crow family, and his books make fascinating reading.

Daphnia (Water Fleas) (See *Live Food*)

A most useful food for all small aquatic reptiles and amphibians, this can be caught in ponds using a fine net, or may be cultured in a variety of food solutions. Green pond water is easiest, or water in which some garden soil, animal manure or cooked liver has been steeped. See *Protozoa* for details of how to culture euglena, a good live food for daphnia.

Deer X

Although these can be seen in semi-domesticated herds in such places as London parks, they are not pets. During "rutting" time in the autumn when the stags are fighting over the females, and again in the spring when the fawns are born, deer can be very aggressive and it is not safe to go near them.

Dew Dispenser

This is a small bottle of water with a tube through the cork in its lid, similar to the drinking bottle used for mice, rabbits, etc. It is suspended upside down in a vivarium containing very small or fastidious creatures, arranged so that water drips on to leaves as a "dew" which the inhabitants can drink.

Dogs *****

Being pack animals by nature, dogs have a strong need for companionship. They like to feel approved of and to be part of an organised way of life. All this puts a great responsibility on the owner, and nobody should take on a puppy without being aware of what is involved. A cat likes to arrange its own affairs, but a dog is totally dependent on its owner for food, exercise and security.

With more than 200 breeds to choose from, the prospective dog owner must think carefully about his way of life. A big dog eats a lot and needs plenty of space and exercise, so will not be suitable for life in a gardenless flat. "Toy" dogs, on the other hand, have not the stamina for an owner who likes to stride across the moors. Pedigree dogs are expensive but will predictably reflect their breed's characteristics; **Labradors**, for instance, will be immensely friendly but the RSPCA say that they tend to rove, and I can certainly think of several Labradors known to me (bitches as well as dogs) which take themselves off for walks occasionally. **Poodles** are intelligent but excitable. Cross-bred puppies, the progeny of two pure-bred parents of different breeds, are usually very strong and healthy and will only vary in type to be like one parent or the other. Mongrel puppies, on the other hand, are of mixed parentage, often with an unknown sire. These may grow up to be anybody's guess where size and appearance are concerned, and although they are usually very robust, a small puppy may easily end up as a very large dog.

Buy direct from the breeder wherever possible, as the behaviour of the mother bitch gives us a clue as to the outlook of her puppies. A friendly, well-fed bitch will pass on her confident attitude to her puppies whereas a badly treated animal cannot give her puppies a good start in life. Ask the breeder what your puppy has been used to eating and at what times of day. A pedigree breeder will usually provide a detailed diet sheet.

Your puppy should be eight weeks old, fully used to eating solid food. He will need four meals a day at first, two of meat and two of milk and cereal such as puppy meal, baby cereal or toasted crumbled wholemeal bread. Raw meat should be cut small enough to be gulped down – over-large chunks may choke an eager puppy. Offal should be cooked but cooked bones should never be given to dogs as they may splinter. Marrow bones are ideal, specially for dogs fed on tinned food, but too many bones can cause constipation.

Puppies seem to eat a huge amount of food, but their rate of growth is very fast and they need to be fed well if they are to develop into strong, healthy dogs. At four months they can have three meals a day, decreasing to two at eight months. Clean water must, of course, always be available. "Convenience" tinned and packeted foods must always be given exactly as the manufacturer's instructions direct. The basis for feeding dogs is two parts of meat to one of cereal, so if a "complete" food already contains cereal, no more should be added. For adult dogs, the cereal is given as dog meal or biscuits, always buying a good brand with added vitamins.

It is important to provide a comfortable, draught-free bed for a puppy and to make sure he or she actually sleeps in it. Many dogs sleep on chairs or sofas but it must be remembered that dogs often have fleas, and fleas breed on bedding, not on dogs. A proper dog-bed with a frequently washed blanket will prevent fleas from becoming a nuisance. "Basket!" later on becomes a useful command if the dog tends to get under your feet.

Right from the start, be gentle but firm with your puppy. Take him outside when he wakes up from sleeping and when he has finished a meal, and wait with him until he urinates or defecates. Praise him and pat him for this as soon as he does it, so that he knows "going outside" is good. Conversely, when he makes a puddle in the house, reprove him gently. Never hit him or frighten him, or you will set up a panic-stricken attitude which will make him much more difficult to train.

Small puppies don't need walks – they just get tired. A securely fenced yard or garden will give them enough exercise at first. Decide whether you want your dog to be with you all the time or not. Adult people who drive cars can often take a dog with them wherever they go, so the puppy can be used to car travel right from the beginning. If you have to go to school, the puppy will have to get used to being without you during the day but he will look forward to a game or, later on, a walk every day when you come home. Dogs sleep a lot provided they are well exercised at least once a day, so the time spent alone passes quickly for a dozing dog.

If there are no humans in the house all day, the dog should have access to a secure garden or yard via a dog-door in the kitchen door, for it is unfair to expect him to remain continent for hours on end. Alternatively, a warm, dry kennel should be provided in the garden. (see diagram)

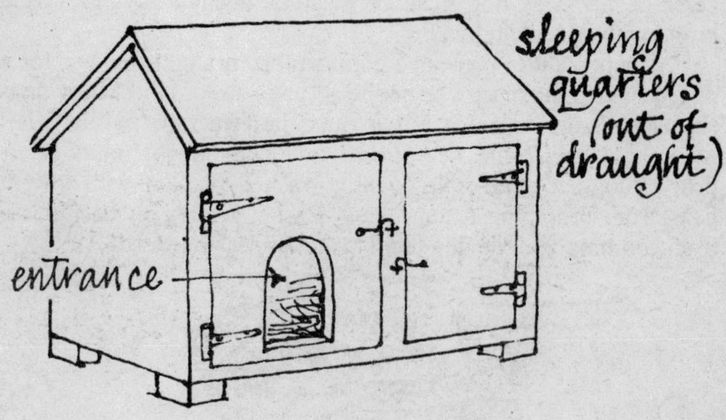

All puppies must be vaccinated against the common – and lethal – dog diseases and until this is done at 9–12 weeks, they should not be taken into towns or allowed to mix with other dogs. This vaccination must be repeated yearly and your vet will give you a record card on which is entered the date of each injection and the vaccine used. This will be required if your dog ever has to go to a boarding kennel if you are on holiday, so keep it carefully.

At six months old your puppy has to be licensed and must wear a collar with a disc bearing his owner's name and address. The licence is quite cheap and comes from the Post Office. If you find a stray dog you are legally obliged to return it to the owner or take if to the police. Anyone involved in a car accident with a dog must also report it to the police. Anyone allowing a dog to attack sheep, cattle, poultry or other farm livestock is liable to prosecution and a farmer may shoot any dog he finds worrying his stock. This is particularly important if you go on holiday in the country, taking your dog with you. Many town-dwelling dogs, specially of the working breeds, get very excited at the sight of farm animals. The untrained instinct to herd them is very strong, and unless your dog is absolutely obedient he may rush after them and do terrible damage. So keep a careful eye on your dog in the country, and put him on the lead rather than take any risks.

Any dogs or other animals coming to Britain from another country have to go into quarantine for six months. Any attempt to evade this law is punishable by a heavy fine, imprisonment and the destruction of the dog – so don't try to smuggle a dog in from a holiday abroad, and don't be tempted to think of taking Fido to France, for you can't bring him back. This is a very necessary rule, for it protects us and our pets from the terrible disease of rabies, which kills both humans and animals in a particularly painful and distressing way. Happily, Britain is so far free of this scourge.

Breeding from a dog should never happen accidentally. Decide at the start whether to get a bitch puppy or a dog, and if a bitch, whether you intend her to breed. If not, you must keep her strictly away from dogs twice a year when she comes into season. She loses a little blood at this time, and is most fertile when this bleeding has just stopped. This, of course, is the time to take her to a dog for mating if you want puppies. Aim at the twelfth day of her season – which should preferably be her second – and allow plenty of time for the mating, as a pair do not always copulate straight away, and when they do, may remain "tied" for half an hour or more. Never attempt to separate dogs paired in this way – just let nature take its course.

Puppies are born about 63 days after mating. Small breeds produce 1–5 young, larger breeds up to twelve or more. Cross-bred bitches are less likely to need any assistance in whelping than pedigree ones, but if a bitch strains continuously for more than an hour without producing a puppy, call your vet. If a bitch has mated with more than one dog she will have a "mixed" litter but this does not detract in any way from the pedigree puppies among them and it does not spoil her for future breeding. If a bitch gets out, to everyone's dismay, take her to your vet within 24 hours of the mating for an injection to prevent conception. Bitches quite often have "Phantom" pregnancies where no developing puppies are present, and if this becomes a continuing problem it is wise to consider spaying. A spayed bitch cannot produce puppies and does not have the hormonal cycle which makes her body constantly prepare for them, since her ovaries and uterus are removed.

A bitch should be wormed before mating and should be free of fleas, lice, mange etc., for these nuisances can be passed on to her puppies. Children handling puppies must always wash their hands afterwards and avoid kissing them, for a parasitic worm, *toxocara canis*, can be transmitted, sometimes causing blindness.

Pregnant bitches must be very well fed, allowing up to three times the normal amount. A whelping box or bed lined with plenty of clean newspaper must be provided, and she is best left alone to have her puppies, the owner looking in occasionally to check that all is well. Never allow strangers to fuss over her, and be prepared for her to be fiercely protective of the new-born puppies. It is rash to allow young children near her at this time. Puppies are solid, heavy little animals, blind for the first ten days but well furred and fast-growing. Finding homes for them is your responsibility – advertise them and put cards in newsagents' windows, and tell your vet about them in case he knows of any good homes.

Obviously there is a lot to learn about the particular breed of dog you choose. There are literally hundreds of books available, covering the detailed management of every kind of dog. Most public libraries have a good selection.

Donkeys *****

Used for carrying and carting for thousands of years, donkeys are now having a sort of holiday (in Britain, at least), being kept purely as pets. They are immensely endearing and intelligent, and are far less expensive to keep than ponies of the same size, being good grazers and economical "doers".

Because donkeys seem naturally fond of people, a single one can be kept without fretting for a companion. However, donkeys are friendly to animals of other species and exert a calming influence on ponies or horses. They help to keep pasture in good condition as they eat the tufts of coarse grass which horses leave, and happily consume brambles, nettles and thistles.

The absolute essentials for donkey-keeping are, first, a paddock of at least half an acre and, second, a sound, draught-proof shelter. Donkeys hate wet as their furry coats soak it up. For this reason they should never be tethered and left unattended. A fence about 1 metre (3′ 6″) high will keep a donkey in as long as it has no little gaps, for donkeys are great fiddlers and pushers and will find any weak spots or easily-opened gate catches. Check that no ragwort (tatty yellow flowers) is growing among the grass and beware of the very poisonous yew tree. Laurel, rhododendron and hot piles of grass clippings are also dangerous.

Donkeys need roughage and do very well on rubbishy grazing. If your paddock is impeccably weed-free, hay must be supplied in a net in his shelter, and your donkey may well eat a lot of his straw bedding. In winter he will need hay at all times (about half a ton should see you through) and should also have some concentrates. This can be in some form of low protein cubes, or a mash of two parts flaked maize to one of bran and one of rolled oats. Sugar pulp nuts are useful but must be well soaked for at least 24 hours before use. Household scraps such as vegetable leaves and stale bread are popular and most donkeys love peppermints, although these are not, of course, good for their teeth. Don't overdo the feeding. Half as much as a pony eats will do. Beware of lush spring grass, for donkeys can get laminitis (a foot inflammation) just as readily as horses. Water must be constantly fresh, the bucket securely tied. Do not regard a field pond as adequate. These get fouled and can harbour all sorts of bugs.

Buy a good donkey. The Donkey Breed Society will advise you (see *Useful Addresses*) or see *Horse and Hound* for advertised stock. There are lots of events for donkey owners to take part in, but a knock-kneed, narrow-chested ass will give you no fun. Good "points" are largely common sense, but any specialised book on donkeys will list them. Try the public library. Donkeys come in many sizes but the big ones over 12 h.h. are known as Spanish donkeys. Grey ones, with black dorsal stripe and the cross over the shoulders attributed by legend to the carrying of Christ to Gethsemane, are normal, but donkeys can also be chocolate, black or cream.

A female ("jenny") donkey comes into season every three weeks for seven days, except perhaps in a cold December and January, and always a week after foaling. She shows little sign of oestrus, seldom being unladylike enough to show sexual excitement by mounting another animal as cows, horses and dogs do, so it is best to run her with a stallion if breeding is wanted. When she is going to accept him she affects a curious opening and shutting of the mouth. The stallion ("jack") must be in the stud book and should be chosen carefully for his good conformation and temperament. The gestation period is usually 12½ months, though donkeys can vary a lot. Foaling is usually easy, and the mare should not be fussed over. If she is in obvious trouble, call your vet at once. A filly should be at least two years old before the first mating. Should the mare have no milk, or in the event of utter disaster, ring the Donkey Foaling Bank (see *Useful Addresses*). Donkey mares are much more willing than horse mares to foster another foal and an orphan foal can, if all else fails, be bottle fed. (see *Orphans*)

Donkeys are easy to train for riding and driving. Their famed obstinacy is due to their intelligent summing-up of what is reasonable and what is not, and it is useless to get cross.

Normally very hardy, damp and draughts are the enemies of donkey health. They can get most of the known horse diseases, one of which, "strangles", is endemic in Ireland, a source of many donkeys. Animals imported direct from Ireland should therefore be quarantined for ten days before mixing with other stock. Donkeys must be wormed regularly and their ever-growing feet must be trimmed by a farrier even if they are not shod. Donkeys are said to live twice as long as horses.

Mules are the offspring of a horse/donkey cross. They are immensely strong and willing but lack the affectionate nature of the donkey. They are unable to breed.

Dormice ***

Though basically wild animals, these tame readily, but must not be kept in the small cages sold for mice. Dormice are climbing animals so their cage must be tall enough to hold branches and growing plants. An outdoor enclosure of fine wire mesh is ideal. Provide a nest box and supply plenty of soft hay and moss for bedding.

The **common dormouse** is mouse-sized but a much brighter brown, with a white tummy and a long bushy tail. The **edible dormouse**, a Chinese delicacy fairly recently introduced to this country and also called the "fat" dormouse, is a much bigger animal, almost squirrel-sized, with a grey coat. Both species hibernate from October to mid-April and must not be disturbed, although some nuts should be left available in case of waking up on a sunny winter day. (see *Hibernation*)

Dormice eat a mixed diet of nuts and grain, fresh and dried fruit, toast or biscuits, carrot, apple and lettuce. They can crack hazelnuts for themselves and like occasional insect food such as mealworms (see *Live Food*). Water must always be available.

The gestation period is three weeks, the litter varying from 4–8. Leave newborn dormice strictly alone – they will come out of the nest at about nineteen days old. Their first coats are grey, turning to the adult brown later. Dormice only live for about two years in captivity, though a record of four years has been known.

Never pick up dormice by the tail – the skin may strip off.

Doves (see *Pigeons*) ****

Doves traditionally live in a dovecote, which is a circular wooden house on a stand, provided with dove-sized entrance holes and perches, and containing a hopper of food which only needs topping up once a week or so, according to size. **White doves** kept in this way are very pretty and trouble-free. They breed prolifically and can become a nuisance in the garden, being closely related to the marauding wood pigeon.

The little **barbary doves** come in a variety of pinky-grey colours and become very tame, so are quite suitable to fly free. **Diamond doves**, with a scarlet diamond-shaped patch round each eye, are better kept in an aviary. Quite small birds, these like millet and other small seeds. Barbaries can be fed as pigeons.

DROSOPHILA (see *Fruit Flies, Live Food*)

Ducks (see *Muscovy Ducks, Poultry*) ****

The most engaging of domestic poultry, ducks look up at you with an intelligent eye and live sociably among themselves. They can be kept with the minimum of fuss; in fact, the **Mallard**, or **wild duck**, is apt to arrive from the sky and live with you as a wild bird does, often interbreeding with tame stock but always maintaining strict independence.

The white **Aylesbury duck** is the one bred for table use, but they also look very decorative on a pond, and become very tame. The smaller **Khaki Campbell** is a prolific egg-layer and the **Muscovy duck** has an entry to itself as some biologists contend that it is really a small goose.

A pond is not essential for duck-keeping, but in hot weather ducks get very despondent without water to bathe in. An old bath sunk in the ground will do, or a moulded plastic pool makes an ideal small pond. A hole lined with polythene is not much good, as ducks are great fiddlers with things and would no doubt tear it all up quite soon. Water is essential if ducks are to breed, for the female duck can't get into the right tail-up position on the ground, and simply gets trodden flat. Drakes have a curly feather above their tails.

Ducks are puddly things, and if you try to keep them in a shed they soon make a sloppy mess of their straw, and gum up their feathers with muck. It is much better and easier to let them loose. They won't go away, and low fences are enough to keep them off the road or a cultivated garden. Make sure netting fences are well pegged down, or the ducks will wriggle underneath.

Ducks are quite keen grazing birds although, like horses, they tend to leave tufts of long grass. They will gobble up any kind of grain and household scraps. New-hatched ducklings should be started on chopped hard-boiled egg, followed by chick crumbs and growers' pellets.

Duck eggs hatch well in incubators (see *Eggs*), provided they are kept fairly damp. Incubator-hatched ducklings are not waterproof, for they lack a mum to rub them over with the oil she secretes under her chin to make them float. It is important to remember this, for ducklings hurl themselves into any water they can find, and often drown. A cold, sodden, half-dead duckling will revive if put back in the incubator or near the hot pipe in the airing cupboard.

Unlike chickens, ducks lay their eggs at night, all over the place. They often lay along the edges of a pond, where the eggs roll into the water, or among nettles where you can't see them or get stung to death reaching them. A well-organised duck-keeper will shut his birds in at night (thus keeping them safe from foxes) and let them out fairly late in the morning – at about nine o'clock, when the eggs will all have been laid.

Ducks tend to be feckless mothers, leading their staggering line of babies about on endless walks until the poor mites die of exhaustion. This kind of idiot mum needs to be penned with her family until the young ones are old enough to stand the pace. She will resent this at first, and hiss a lot, but plenty of good food will soon persuade her that she has made a wise decision.

Although usually healthy, ducks can suffer from a form of progressive paralysis, sometimes arising from damage to the feet on rough ground. Antibiotics help – ask your vet.

EAGLES (see *Birds of Prey*) X

Earthworms (see *Live Food*) **

To see the amazing effect earthworms have on the soil, it is interesting to arrange layers of different-coloured soil ingredients such as sand, compost, chalk and loam in a big glass jar, topping the layers with some rotten leaves. Introduce a few earthworms and watch what happens. In a few days the layers will be giving way to an even mixture, the leaves drawn down to enrich the soil.

To breed earthworms as a live food, a larger container such as a barrel or wooden box is best, filled with moist earth (not heavy clay) and leaf mould. The container should have small (worm-proof) holes or cracks in the bottom for drainage, and must be covered with sacking and kept out of the sun, for a temperature of more than 16°C (62°F) may be fatal. Earthworms breed freely in good conditions, mating and laying eggs which form as a "saddle" on the female's body, later deposited as a cocoon. If plenty of rotting leaves are supplied, the worms will need no extra food, but they can be given some oatmeal soaked in milk as a treat.

Earwigs **

Earwigs can be kept in a big jar or a sandwich box, the bottom well covered with sand. The male has bigger pincers on his rear end than the female, and they mate in early spring. Eggs are laid in a hole in the sand (or earth or rotten wood in nature) and are guarded quite fiercely by mum. Earwigs are fairly omnivorous but have a preference for a meat-based diet. Try chopped mealworms, maggots, aphids etc. as well as greenstuff, bran, breadcrumbs and sieved hard-boiled egg.

EGGS (to hatch) (see *Amphibians, Birds, Fish, Poultry, Reptiles*) ***

All babies come from eggs, although in most of the animals we know best the egg is a tiny thing fertilised inside the mother's body by the male and growing into a baby which is born alive when it is big enough. There are lots of other systems. Eggs can be laid ready fertilised, as in the case of frog spawn, or they can be laid and then fertilised externally – many fish do this. Some insects keep their eggs inside their bodies until they hatch, thus producing live young, and a few oddities such as stick insects manage without any fertilisation at all.

The eggs most familiar to us are ordinary hen's eggs – and yet they are not ordinary at all. Most birds mate, then lay the number of fertile eggs they intend to hatch. They they stop laying, incubate their clutch of eggs and rear their babies. No so with hens. When a young pullet reaches about 21 weeks old she starts to lay eggs, regardless of whether or not she has mated with a cockerel. Even if she never mates, she goes on laying eggs for most of her life, though in decreasing numbers, and stopping each year when she moults her feathers. Such eggs will never hatch, for they are not fertile. If a cockerel runs with the hens he will mate with them and the eggs they lay will then be fertile, and will hatch if a hen goes "broody" – which means she develops an overwhelming instinct to sit on a clutch of eggs until they hatch out as chicks. These eggs may be her own or another hen's, or even the eggs of a duck or turkey or pheasant.

The poultry-keeper whose hens run with a cockerel can hatch eggs when he chooses, provided he can keep them as warm and as moist as they would be under a hen's feathers. An incubator, running on paraffin, electricity or gas, will provide these conditions and commercial hatcheries produce millions of chicks with never a broody hen in sight.

It's no use trying to hatch half a dozen eggs bought from a shop, because they are probably produced in a "battery" consisting of caged virgin hens. Eggs from a back-yard poultry-keeper who runs a cockerel with his birds can be hatched in a small incubator bought at moderate cost or even improvised from a wooden box and a light bulb. (see diagram)

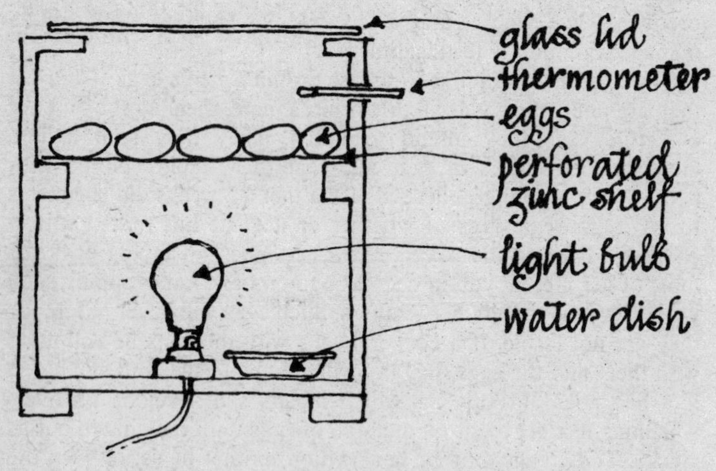

38°C (104°F) is the temperature for incubating eggs and ideally the incubator should have a thermostat to keep this temperature steady. A dish or tray of water must be kept full, as the air should be rather humid, as it would be under a slightly sweaty hen, The eggs must be turned at least twice a day, preferably more, to keep the developing chick from sticking to the side of the shell, but in the last week it is best to leave the eggs unturned so that the embryo settles in a properly orientated way, beak up. Hen eggs hatch at 21 days but you can hear them cheeping a day or two before. Don't try to help a chick out of its shell, or you will probably rupture a blood vessel and kill it. Not all eggs hatch, and there are likely to be a few "dead in shell". Don't grieve – it is nature's way of weeding out weak or deformed chicks.

New-hatched chicks need nothing to eat for the first day as the yolk-sac they absorbed in the egg will supply them with enough. As day-olds they should be strong and chirping, and well dried off. They can be transferred to a brooder which gives them warmth and space to run about, and the best food to start them on is crumbled hard-boiled egg. Gradually mix in some commercial chick crumbs from a miller or garden and pet shop, and supply water in a spill-proof shallow dish which they can't drown in. A jam jar upside down in a saucer will do, but strong chicks tend to knock it over, so I prefer a lemon squeezer which provides a safe moat of water. Brooding temperature should be 32–35°C (about 100°F) at the heat source (a black-painted light bulb is easiest), allowing the chicks free access to an unheated run. At 6–8 weeks, depending on the weather, the chicks can do without heat as they will now be well-feathered little birds.

Bantam eggs hatch in as little as 18 days, ducks, geese and turkeys in 28, guinea fowls in 24. Young water birds have very wet droppings and like to live in a puddly state, so it is best to house them separately from chicks, which need to be kept warm and dry. Even the eggs of ducks and geese like to be damp, and their hatching is much improved if you turn them with a well-wetted hand each day.

Baby birds are convinced that their mother is the first large thing they see, so hand-reared chicks think of their owner as mum. This is endearing, but unless you are firm they will all come into the kitchen with you, so watch out!

ERMINE (see *Stoats*) X

FAWNS (see *Deer*) X

Ferrets ***

Ferrets are descended from the polecat family – in fact, non-albino ferrets are commonly called polecat-ferrets. They are mammals, have broad heads, sharp teeth and long bodies with the humpback characteristic of wild stoats and weasels. Through in-breeding they are often quite small but can reach a body length of 45cm (17½″) for a dog, the bitch smaller. The tail can be anything between 13–19cm (3–7″) long in addition to this. Colour varies a lot through cream to darkish brown with pale yellow underbelly. Pink-eyed albinos are common. They are not noisy animals, but have a chattering alarm call and will also hiss aggressively if upset.

Cages for ferrets must be very strong. Wire netting is no good – they bite through it. Welded mesh is the only suitable material, and it is best to use this for the floor of the cage as well as the sides, so that droppings fall through and can be cleaned up easily. A solid floor soon soaks with urine and develops a fearful smell. The nesting quarters can be solid as no ferret would dream of fouling its bedroom. This should be well filled with hay or straw. (see diagram)

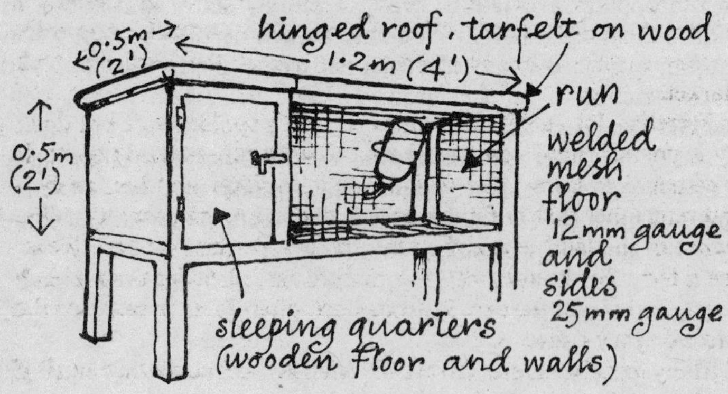

Ferrets must have meat every day and if you can't get it or can't afford it, don't keep ferrets. Cull chicks from a hatchery are the easiest diet, two or three chicks per ferret per day. They like bread and milk but it will not do as a staple diet. Many owners leave their ferrets unfed one day in seven to ensure that all stomach contents are totally digested. This reproduces the natural state of things where a hunting animal will occasionally be unlucky and fail to kill itself a meal. It also keeps the appetite sharp and avoids the dangers of obesity. Many zoos adopt this procedure with their lions and other carnivores.

Breeding starts in the early spring and a bitch ferret may come into season from February onwards. You can see this by inspecting her rear end; when in season her vulva will be swollen, protruding in a little pea-sized bulge. Mating is a noisy, aggressive affair but don't worry – they like it that way. The young will be born in 41–42 days, small, blind and bald, and up to six in a litter. Ferret mums are very touchy, so leave the babies strictly alone and of course make sure the bitch has a cage to herself in which to produce her family. She won't take kindly to sharing! Be lavishly generous with food for a lactating bitch and growing young ones. They can be weaned at eight weeks, when they can leave their mother, and at this stage they need lots of stroking and cuddling and handling so that they will grow up tame and friendly. The bitch may possibly produce a second litter in the same year if the first lot were born early, so watch her for coming into season again. Before deciding to breed ferrets, though, make sure you can sell the young ones, or you will fetch up running a large and smelly colony which is more than you can cope with. Poultry dealers often buy ferrets, or ask at the local livestock market.

Ferreting for rabbits continues to be a popular sport but don't try it yourself until you have been out with experienced people. It is essential to know how to kill rabbits quickly and humanely – and remember that in England wild rabbits are the property of the owner of the land where they live, so ask permission first. Never use a ferret for hunting rats – a mature rat is heavier than a bitch ferret, and much nastier. Rat bites are often fatal because of the infection they cause.

Many people regard ferrets as offensive because they smell if not kept scrupulously clean, and their bite can be dangerous. Like any other pet, a ferret is as good as the way it is kept. Half-starved, roughly-handled ferrets are a menace but if properly looked after they can be affectionate, rewarding pets.

FINCHES (see *Cage Birds* and individual headings for *Bullfinches, Zebra Finches*) ***

FISH (see *Goldfish, Minnows, Sticklebacks, Tropical Fish*)
Fish to keep as pets fall into three categories: coldwater species such as goldfish; seawater fish and other marine life, and tropical fish.

Water-dwelling creatures have feelings just like anything else. They need a good environment to live in, as near their natural conditions as possible, and the prime essential is an adequate supply of oxygen in the water. This does not necessarily mean that an aerating pump is essential – water takes in oxygen from contact with the air and a wide aquarium tank will supply itself with oxygen provided the fish are not overcrowded. As a rough guide, allow 2·5cm (1″) of fish (not counting tail) to every 155 sq.cm (24 sq.ins) of water surface. Even more roughly, but simply, see how many times you could put your hand flat on the surface of your tank, not overlapping. This is the number of fish-centimetres you can have. If, say, it works out at 30 fish-centimetres, you could have three 10cm (4″) fish or two 15cm (6″) fish or twelve tinies. It doesn't matter how you divide it up as long as you don't overstock. When the fish grow bigger, some of them must be moved to another tank, or they will start to come to the surface of the water, trying to gulp air – a sure sign of oxygen starvation. The little round bowls so often seen at fairs are hopelessly low in oxygen because of their narrow necks, particularly if they are filled to the top. As the diagram shows, they have a wider water area if only filled half full – but they should never be used to house a fish permanently.

small oxygen area

larger oxygen area

Aquarium tanks have to withstand a lot of pressure, for water in bulk is heavy stuff. Get a good, strong one and fill it for the first time in the kitchen sink or outside if it is too big. This will show you any leaks without flooding the sitting-room carpet. Aquarium cement will solve this problem.

Don't rush out and buy fish until their home is completely ready. See *Tropical Fish* for notes on setting up a heated tank, and *Sea Creatures* for a marine tank. For ordinary goldfish, shubunkins etc., tap water at room temperature is fine, but it should be stood in a bucket for twelve hours in the room where the aquarium is to be set up. This enables it to assume room temperature and rids it of the inevitable chlorine. First, though, the aquarium must be carefully bedded with aquarium compost, thoroughly washed to free it of dust. Pile this higher at the back of the tank, sloping down gently to under an inch (2cm) at the front. Add any rocks you fancy, taking care not to make any wobbly structures which may collapse on the fish, and avoiding any sharp edges. Natural stone from garden or seaside should be boiled before use to kill any harmful micro-organisms. Cover the compost layer with aquarium gravel or chips. Now fill your tank half full with water from your bucket, pouring it very gently into a cup (see diagram) so that it does not wash your gravel about.

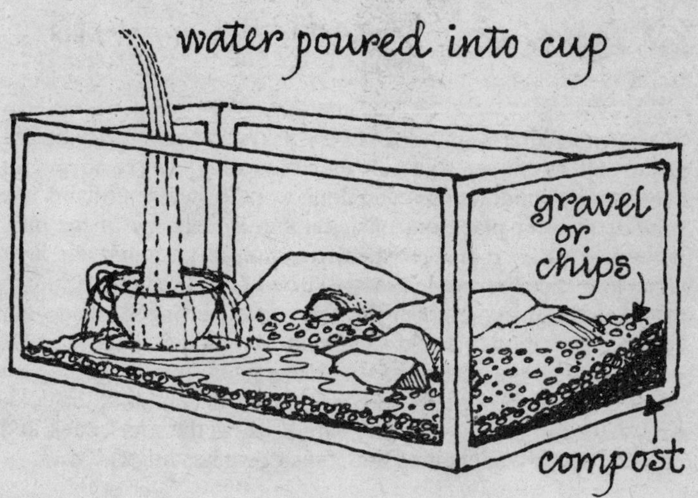

Now for the aquatic plants. These are vital to help supply oxygen and to keep the tank clean, for they live on the manure produced by the fish. There are heaps of aquatic plants but some of the prettiest ones live in warm water, so check carefully when you buy them that they are suitable for a cold-water tank. Some of the commonest kinds are tape grass (vallisneria), Canadian water weed (*elodea*), and the water violet (*hottonia*), with fern-like leaves. Plant them in the compost, spreading out their roots carefully and firming the compost and gravel back over them. A small strip of lead can be fixed like a collar round the base of the stem if the plant tends to float up. Now top up the tank until the water is level with the bottom of the top strip of angle iron. Leave it for a few days to settle down – any greenness will clear itself quite soon.

Remember that a full tank is very heavy. Once in place, it will be difficult to move, so think carefully about where to put it. Avoid sunny windowsills – you don't want to stew your fish alive. A moderately lit place out of direct sun is ideal, but if the only convenient place is a darkish corner, rig up an aquarium light with a long metal reflector over it.

Choose bright young fish for your tank, bearing in mind your stocking limitations. Avoid any with the slightest suspicion of furriness round the gills or chewed-looking fins or tails.

See *Goldfish* for various breeds and bear in mind that the cold-water tank can also house such native fish as the stickleback and minnow, which you can find under their own headings.

Fish are not big eaters and overfeeding simply fouls the water. Only give as much as the fish finish up in five minutes. They like variety and if proprietary dried food is used, change the brand from time to time. Live food such as daphnia, mosquito larvae, microworms or tubifex (sludge) worms have the great advantage of not going "off" before they are used, and fish love them. (see *Live Food)*

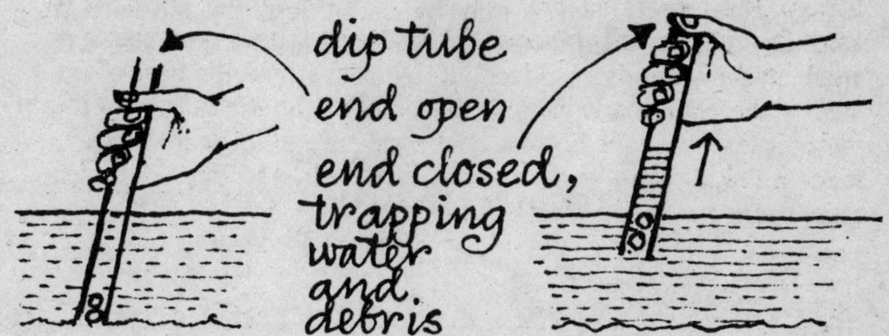

Tanks must be kept clean and free of rotting food. A siphon tube or a glass "dip tube" (see diagram) is essential for picking up any debris from the bottom, for an accumulation will result in the smelly "black sand" which poisons the environment. In general, fish remain as healthy as the water they live in. If the tank begins to leak or through neglect becomes really messy, a complete clean-out is necessary. Here a spare tank is almost essential, filled with water at the same temperature as the other tank. Move the fish by using a soft, fine-meshed net or, easier, two nets so as to trap the fish between them. Never handle fish, or you are likely to rub off part of the protective slimy coat, admitting infections.

Well-kept fish will breed, the female laying eggs which are fertilised afterwards by the male spraying them with "milt" from his body. The young hatch in about four days as thread-like creatures with the egg-sac still attached to them, supplying all their food needs until it is totally absorbed. Parent and other fish regard these youngsters as live food, so if the owner wants to preserve them he must move them to a separate tank with plenty of aquatic plants. A better method is to put a likely-looking male and female into a separate tank, returing them to the main tank when the eggs have been laid and fertilised.

Young fish are best fed on the protozoa living in pond water, which is one reason why fish appear to breed so well in outdoor ponds. For tank use this tiny animal life can be cultured very simply by infusing some chopped lettuce in some well-established aquarium water. In a few days the water will look cloudy due to the presence of millions of organisms, and a spoonful of this makes a meal for about fifty fish "fry". The female may lay 3,000 eggs, a thousand of which may hatch, so don't try too hard to rear the lot. Scraped liver and sieved hard-boiled eggs feed fish.

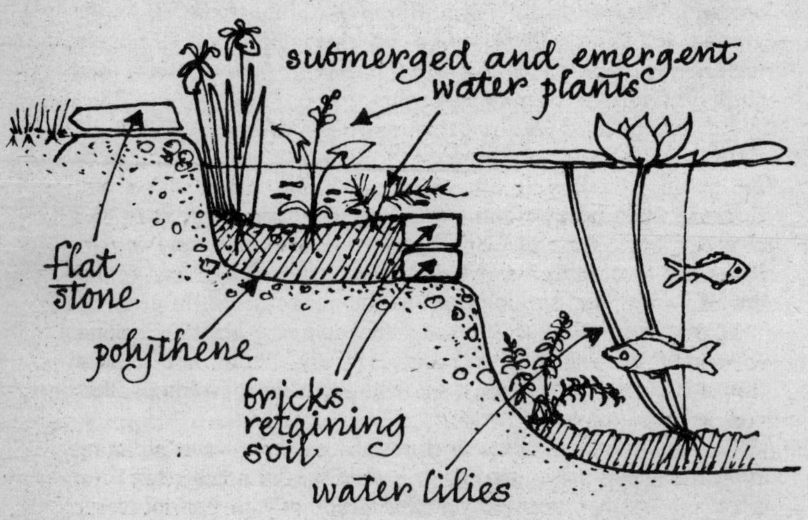

A garden pond, if well built and kept, is attractive to look at and provides a very natural habitat for fish. A hole dug in the ground and lined with 500-gauge polythene is a cheaply made pool, or a pre-formed but more expensive fibreglass mould can be used. Either should have a shallow "step" at the edge for aquatic plants to be established, for this is where the fish will like to breed. Always check carefully that the original hole in the ground has no sharp projecting stones or splinters which may cut the polythene.

Water lilies can be grown in the deep middle of such a pool, providing valuable shade for the fish. Ask your supplier what depth of water these need, and tell him the size of your pool, for some varieties need a Kew Gardens-sized lake. Let the pool settle down for some time before introducing fish, and bear in mind that the less hardy species are not suitable for outdoor life. The **shubunkin** is very robust and will mix happily with **carp, golden orfe, rudd, tench** and **minnows.** Avoid the pugnacious **stickleback.** Snails are not necessary and they breed so fast as to become a menace.

Outdoor fish are not fed in the winter. Start feeding in the spring, giving earthworms once a week. (A damp sack left on the ground will always have a few on the underside.) Gradually increase the feeding rate to twice a day in the mid summer, then slack off again as autumn approaches.

Remove fallen leaves and floating rubbish from the pond and keep the water plants in check by occasional pruning, cutting out the oldest growth near the main stem. Lilies can be lifted and divided in the spring. Once a year the pool must be emptied and cleaned, first moving all the fish to a temporary tank. Do this, too, if fish are gasping at the top of the water – an indication of over crowding or fouled environment.

"Cotton-wool fungus" on fish is cured by moving the sufferer to an isolation tank with one tablespoonful of salt to every 5 litres (about a gallon) of water. Keep him there until the fungus has completely disappeared.

Ice on outdoor pools in winter should not be broken by hitting it, as the shock wave may damage the fish. Put a hot kettle on it until a circle has thawed. Do this every day to provide some oxygen.

Fishing cats can be foiled by having a wide, wet but fish-free bog garden round the pool, stocked with water-loving plants. Few cats will get all four feet wet, even for fresh fish.

Flatworms *

Mainly interesting because, if cut in half, both halves of a flatworm (or "planarian") will grow into a new worm. They can be found under stones or on the undersides of water plants. They hate daylight and should be kept in shallow dishes of pond water, partly covered to provide darkness. Feed every 4–5 days on pieces of raw liver which must be removed a few hours later, and the water changed.

FOALS (see *Donkeys, Horses and Ponies*) *****

Foxes X

Although you may see people on television cuddling fox cubs and other wild things, it's not as easy as it looks. A fox can only be tamed if it is taken as a very young cub, and even then the life it leads will be an unnatural one, deprived of the company of its own kind and the chance to breed. Foxes in the wild eat such things as mice, frogs, small birds and insects, and it is a sure sign of the man-made damage being done to the environment that foxes are spreading into towns and can even be seen rootling in dustbins. Being intelligent, easily adaptable creatures, they may possibly "tame" themselves at some future time, as dogs must have done centuries ago – but the initiative must be left to them.

Frogs **

(*Tree Frogs* ***)

Because of the grim process known as civilisation, the wild, dabbly places beloved of children with nets and jamjars are rapidly disappearing. The reedy patches of wasteland where dragonflies flitted are now car parks and the ponds where the frogs bred have gone. And the frog, once the most common of British amphibians, is struggling for survival. Because of this, frog spawn, the jelly-coated eggs which frogs lay in huge quantities in March and April (but at no other time) must not be taken for granted as a free springtime amusement. Unless you are serious about rearing some little frogs, leave it where it is. And if you really want to watch some tadpoles develop, just take a little spawn – about a tablespoonful is quite enough, separated from the big mass.

Keep your spawn in pond water, not tap water, and make sure there is a little mud from the pond at the bottom of the container. This will provide minute forms of life which the new-hatched tadpoles can eat. (The word "tadpole", incidentally, is a corruption of "tailed poll", poll meaning head. So, literally, it means "a tail-head".) If you live at some distance from the pond, make sure you take a supply of pond water in a big jar or bottle so that you can top up your tadpoles if necessary. The black dot in the centre of each egg is the embryo tadpole and you can see this become comma-shaped and begin to twitch. It lives at first on the "jelly" surrounding it, then breaks out to live independently like a little fish. Its gills give way to an internal breathing system and it grows first hind legs then front ones, suddenly looking very much like a tiny frog with a tail. At this stage you must transfer your tadpoles to a wide container with some flat stones sticking up from the water, for the little frogs need to breathe air. They can also jump, so make sure your container is deep enough to hold them or cover the top with muslin or fine net. And watch out for cats.

Tadpoles in the "fish" stage will eat each other if they are not supplied with enough protein. A small piece of meat suspended in the water on a piece of string will give them a good source of food to suck at, but you must change it frequently. Sniff at your tadpoles each day. If they smell like pond water, all is well, but the minute they smell rotten, you must change the water. Have a new container of pond water ready (and at the same temperature), then simply tip the old container away through a nylon sieve (beware of old metal sieves with sharp bits of wire). Quickly remove any bits of old meat which have slipped off the string, then tip the tadpoles into their new container. At the little frog stage, the diet changes, for adult frogs eat flying insects. Don't offer dead insects, for frogs are only stimulated to eat by the sight of moving prey. Greenfly are the best thing for tiny frogs, but do make sure Dad hasn't been spraying the roses with insecticide. Things start to get tricky at this stage, and it is time your frogs went home. If you liberate them in your garden some of them may survive and grow but they will never breed there. One cause of the frog's downfall has been his insistence on returning to his native pond to breed – bad luck if it has been turned into an underground pipe. Take your little frogs back to the pond where you found the spawn, and turn them loose there. Out of the 2,000 eggs laid by the mother frog, only six on average will reach maturity left to themselves, so if you have improved on that figure, you have done well.

There are about 200 species of frog, some of which are suitable for keeping as vivarium pets. Our **common frog** is not included in this list because of its diminishing numbers and because it never breeds in captivity. You may, of course, have frogs living naturally in your garden if you have a pond, and they will do well providing you leave their surroundings damp and shady. An energetic clipped-lawns-and-gravel-paths gardener can ruin things for frogs. The common frog is about 7cm (3″) long, brown, yellow or olive green, and hibernates in the winter. We also have the **edible frog** which is rather bigger, with black spots and yellow stripes down the back, with cream underparts. These are a delicacy in France and potentially here too – hence their introduction. They can be kept in a vivarium but are cannibals, so keep large ones separate from small.

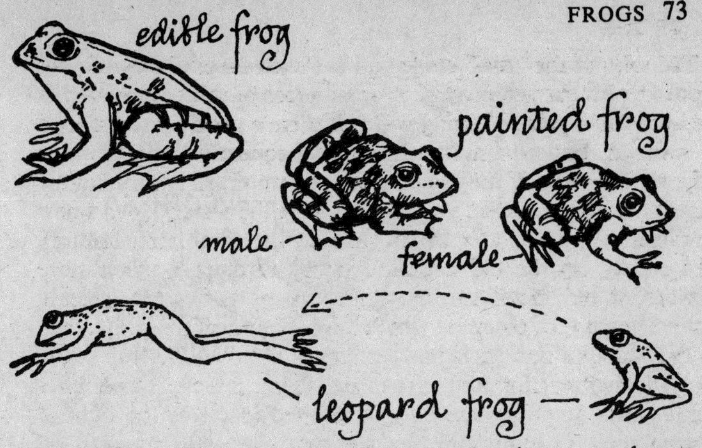

The **painted frog** has dark blotches and stripes on a brown background and catches its prey by a sideways grab with the mouth, not by flicking out a sticky tongue as most frogs do. You can sex these with certainty because the male has webbed feet and the female hasn't.

The **leopard frog** is quite small, 6–8cm ($2\frac{1}{2}$–3″) but can jump nearly 2 metres (6′). It is grey, green or light brown with rows of dark spots and white underparts. It has a diet of worms, grasshoppers, beetles, slugs etc. and the male is reputed to "sing" when picked up (it is probably shrieking with fright).

Beware of the **American bullfrog.** It is the size of a guinea pig, can't be kept in a vivarium but will live happily in a greenhouse with a pond. It eats a huge amount of fish, worms and slugs and constantly bellows for more. It is absolutely insatiable, hideously noisy and will make you very unpopular with the neighbours.

Tree Frogs are much jollier vivarium animals than the ordinary frogs, for they are equipped with sticky feet which enable them to climb up glass. They are also perfectly at home in trees, so their vivarium should be a tall one, containing tree branches or growing twiggy shrubs. They all eat flies and insects, lay spawn in water and develop through the usual tadpole stages.

Only 5cm (2″) long the **European tree frog** is bright green, varying a bit with the colour of the leaves he is sitting on, and yellow underneath. He has a narrow black stripe on both sides from his mouth to his tail, and there is also a very rare blue variant. These tree frogs are reputed to do a kind of "weather house" act, sitting on high branches in fine weather, lower down if cloudy and at the bottom when raining.

The American tree frog is a warty-skinned 7cm (3″) long animal. It is basically brown or pale grey but changes colour to fit its background.

The blacksmith tree frog is so called because its croak sounds like someone hitting an anvil. Green to brown in colour, this is the biggest tree frog, about 13cm (5″) long.

Fruit Flies (Drosophila) (see *Live Food*)
Fruit flies (drosophila are immensely useful for feeding young creatures of many species, and for lizards. These small flies feed on yeast, which is why they gather round squashy fruit. Over-ripe banana in a big jar will provide a home and food, best kept at 18–26°C (66–77°F). Adult flies lay eggs from one day old to their death in about three weeks. The eggs hatch out quickly and the larvae eat voraciously, becoming adult in 9–14 days after two moults. The life span is 26 days for females, 33 days for males, and they are very prolific. For laboratory conditions, drosophila are best fed on a culture of maize meal and black treacle, incorporating a mould inhibitor.

Gecko (Fanfooted Lizard) (see *Lizards*) ***
About 18cm (7″) long, green to olive-brown in colour, quick moving and boggle-eyed, this lizard is remarkable for its funny feet. Its toes end in suckers which enable it to run up walls and across ceilings – in fact, it likes to sleep upside down, hanging from its feet on the roof of its vivarium. Although subtropical in origin, it will be quite happy at normal room temperature provided this does not drop at night. No house-dwelling lizard will like the central heating being turned off. The gecko eats beetles, insects, worms etc. but is nocturnal, hunting its dinner at night. A pair will breed quite readily, particularly if given free range in a warm greenhouse, but lay only one egg. What's more, when their single baby hatches out, they will at once eat it unless it is moved to safety by the wise owner. Quite a fun pet unless you are prejudiced about cannibalism.

Geese (see *Poultry*) ****

Geese are easy to keep, needing no housing. They are very hardy and, unless caught while incubating eggs, are a match for a fox. They are grazing birds essentially and will keep grass neatly mown and copiously fertilised. (Think of about sixteen geese to the acre if you don't want to feed them any meal; the same as four sheep or one cow.) Most people give their geese some wheat or barley twice a day. Bear in mind that geese are vegetarians, so don't give too many high-protein pellets which may consist largely of fish meal. Geese also like apples, carrots, potatoes, lettuce and any other greenstuff. They must have an ample supply of clean water and love to have access to a pond so that they can swim. Deep water is almost essential for successfuful mating. (see *Ducks*)

Eggs are laid every other day from early February to autumn, in an untidy straw nest covering the eggs from sight. Geese are fiercely protective mothers and very ferocious as broodies. If you want to add extra eggs to a broody goose's nest, get someone to distract her attention while you creep up behind her, armed with eggs, and grab her by the tail. Once up-ended she cannot attack you, but retreat quickly once you have pushed the extra eggs under her! Eggs hatch in 28 days but often incubation has started before all the eggs are laid, so that goslings hatch off at two-daily intervals. In this case, the gander often takes charge of the first-born, leading them to food and water and returning them to the still-sitting goose for brooding when they get chilly.

The biggest geese are grey and white or pure white, a complex Anglo-French cross known generally as **Embden-Toulouse.** Smaller but more beautiful and with much more personality is the **Chinese,** with a chocolate stripe down the back of the neck and an orange knob at the base of the beak. The Chinese is a relatively small bird and quickly becomes part of the family since it takes an active interest in whatever goes on.

Geese are even more inclined than ducks to fiddle with things. They fiddle with the wires underneath cars, thus extinguishing the headlights, and fiddle with any loose plaster round windows, sometimes managing to make a hole right through so that they can watch television with the rest of the family. They are also very noisy, specially when hungry. And they are hungry every time they see you.

Canada geese, with black necks, live in wild flocks but are also well-established as domesticated birds and are remarkably tame. All geese make good watchdogs, for they dislike the approach of strangers and hiss fiercely enough to repel unwanted visitors – or even wanted ones. Goose-keeping is full of such disadvantages and yet, having once kept geese, it would seem odd to be without them.

Gerbils

*****

Although unknown in Europe before 1964, this newcomer to the pet scene is a huge success. Being a desert-dweller, the Mongolian gerbil conserves its body fluids, so producing very little urine and small, dry droppings. It is therefore very clean and smell-free.

Black, white and cinnamon mutants exist but the normal gerbil is sandy-brown with a pale tummy. A black stripe runs down spine and tail. The hind legs are long, with five-clawed feet used for digging and burrowing, for gerbils in the wild excavate galleries and breeding chambers deep in the desert sand where they are protected from the burning sun. They are active animals, adapted to busy foraging in order to find food in their sparse desert surroundings. Bearing this in mind, it is clear that small cages are utterly unsuitable for housing gerbils. Although they need a small, private nest-box, they must have big accommodation with room for a good depth of peat or compost in which to burrow, and enough headroom to permit jumping. An aquarium tank is ideal – sometimes non-waterproof ones can be bought cheaply. It will need a lid of fine wire mesh and the water bottle must be fixed fairly high so that the gerbils do not pile litter underneath it, causing leakage. (see drawing)

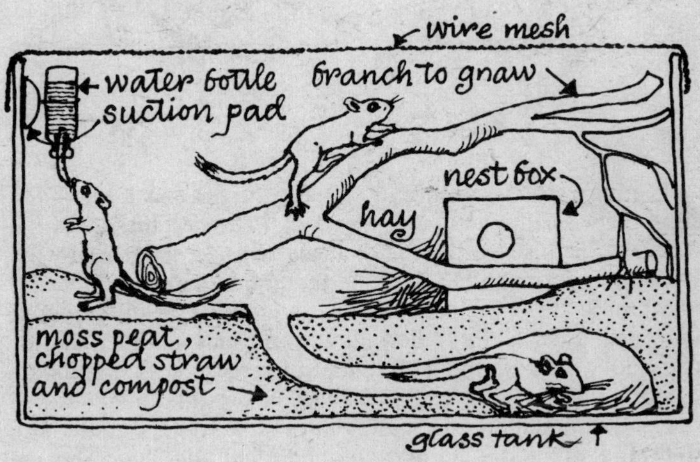

Because their native terrain will not support big colonies of gerbils – or anything else – it is natural to gerbils to fight fiercely for living space. A mated pair will live together for life but must be introduced when very young, or they will fight. Two females may be kept together if breeding is not wanted but, again, must be started together as youngsters. Natural colonies can sometimes be established but only in very large accommodation where each pair can have its own territory.

Feed gerbils on whole oats, wheat, maize, barley, canary seed, sunflower seed, fresh fruit and vegetables, hay, dog biscuits and hard-boiled egg, giving about a tablespoonful of a selection of these per day to an adult gerbil. Allow more for pregnant and lactating females, and offer powdered or liquid milk. Always provide something to gnaw, such as bark-covered wood. Occasional live insects may be liked.

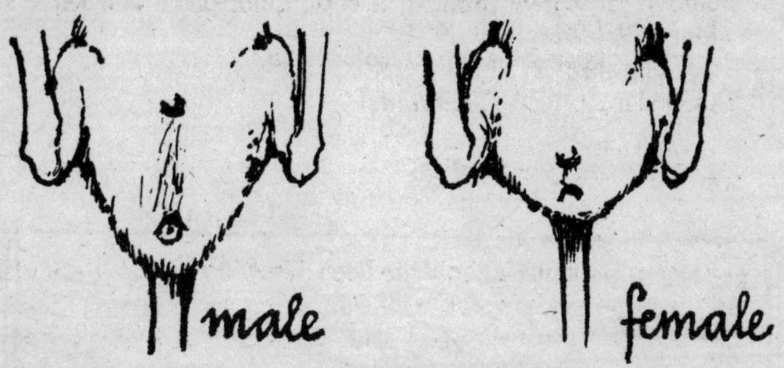

The male's body is tapered to meet the tail and has a wider space between the anus and the genital opening than does the female's. She is shaped more like an ice lolly (see diagram). By ten weeks old, young gerbils must be sorted into males and females and kept in these groups, or paired as breeding couples. Choose unrelated animals for breeding, or problems of infertility or deformed young will arise. A female comes into season every six days throughout the year, remaining in oestrus for about five hours. The gestation period is 24 days on average, the litter size 4–6. The young are born naked and blind and should not be touched, or the mother may eat them. Their eyes will open between 16 and 20 days, by which time the doe may be pregnant. In this case, remove the young to a separate cage when they are 22 days old, to give mum a breathing space before she has the next litter two days later. Gerbils, in fact, control their population by the process of delayed implantation when more than two babies are being suckled. This means the mother does not become pregnant again for two months rather than one. Some gerbils only have three or four litters in their whole life span of 24–36 months, and at best – or worst – it is seldom more than ten.

Gerbils are busy and inquisitive, appreciating a toy such as a cardboard roll or small box. They like shredding paper for bedding, though hay should also be provided. They are usually healthy little creatures but may suffer from heat exhaustion if left in direct sunshine, which can heat up an aquarium tank like an oven. Over-handling can also produce a form of epileptic fits. In either of these cases, leave the gerbil in its normal housing but remove the whole thing to a cool, quiet place and leave strictly alone to recover.

GIANT SPIDERS (see *Tarantulas*) *

GLASS SNAKES (see *Snakes*) ****

Goats ****

Goats are difficult animals to keep. They are choosy, suicidal, and as tolerant of damp as a packet of crisps. They are intelligent, amusing and affectionate, and devoted goat-keepers find them totally satisfying. A goat-keeper is very nearly a farmer, for he is involved in daily feeding, milking and mucking-out, and in growing or buying quite bulky supplies for his stock. No goat is simply an animated lawn mower and all goats involve a lot of daily work, so think carefully before taking one on.

Grazing is not essential, for goats will browse along hedges or rough ground rather than eat short grass. A sound, waterproof, goat-proof shelter is absolutely essential, though, for goats must be kept warm and dry since their coats do not shed water as those of horses and cattle do. A securely fenced yard or paddock is needed for exercise, but goats can jump over a metre (5′) and, because they stand on their hind legs, will break wire netting down with their front feet, often entangling themselves. Chain link is the only suitable fencing apart from closely spaced post and rail, wired at the bottom to retain kids. Goats can hang themselves if tethered.

Before buying a goat – or preferably two goats, as they need company – join your local goat club (see *Useful Addresses*) and get to know some experienced goat-keepers who will show you how to set about things and sell you suitable stock. Goats breed in the spring, having a gestation period of five months dating from a limited period when the females can come into season, from early September to the end of March. This means that kids are born from the end of January to August in theory, most of them arriving in February, March, April and May. A goat can have as many as five offspring from one mating but singles, twins or triplets are more common.

Most goat-keepers retain only the female kids, humanely destroying the males at birth. Male goats are very large and strong and in the mating season they reek. One male is enough to serve a lot of females for they are randy animals and a male can beget kids at only a few months old himself. Obviously, only a few very well-bred males are needed as stock animals. Surplus male kids can be reared for meat and some people keep castrated male goats as hedge-trimming and lawn-mowing pets. These can be very affectionate and like to be taken for walks but all too often their owners get tired of the constant work of providing for them, specially in the winter when the lawn does not grow and the hedges are bare. Hay and cereals are expensive, and a full-grown goat eats a lot.

As a general rule, the intending goat-keeper aims at producing milk, either for the house or to feed other livestock. Goats' milk is ideal for feeding puppies or piglets or orphaned young of any species (see *Orphans*). Having very small fat globules, it is like homogenised milk, slow to separate its cream. It contains no yellow colouring so always appears pure white, as is butter made from it.

Start, then, ideally with two female goatlings, which are kids over a year old and ready to mate in the following autumn. Pedigree stock or at least animals of a recognisable type are better value than "scrub" goats as they will milk better, look nicer and bear more valuable kids. Keep them together or in adjacent pens where they can see each other, and feed them hay in a rack, for they will not eat any which spills on the floor, and a mixture of rolled cereals, bran and flaked maize. Many millers stock a goat mixture, which saves time in mixing. Root vegetables such as swede, mangels or fodder beet are liked, as are greens of all kinds – but make sure these are free from water or frost. Drinking water must be changed at each feed, the bucket securely tied. Leafy branches and hedge trimmings are a favourite food, but avoid the poisonous laurel and rhododendron (if in doubt, don't give anything with glossy leaves).

Goats come into season every three weeks for two days, during which time they often bleat incessantly, wagging their tails agitatedly. Some keep rather quiet about it and with these you have to look for the slightly swollen vulva and trace of clear discharge. Take the female to the male and hold both of them firmly by the collar and/or beard until the actual mating takes place, or madam may rush off in a panic with sir in hot pursuit.

Kidding is usually easy but when several babies are present their heads and legs can get tangled, resulting in an awkward "presentation". Call your vet at once if your goat strains for a long time without producing a kid, and don't try to help her yourself. Random pulling can do a lot of damage. Other veterinary jobs include disbudding kids at three days old, to prevent their long and possibly dangerous horns from growing, and giving routine injections against entero-toxaemia, the most lethal of goat diseases. Worming and foot-trimming can be carried out by the owner, but ask an experienced goat-keeper how to set about it first.

Kids must be left to suckle from their mothers for at least three days so that they have adequate colostrum, or first milk. This gives them protection from infection and a big nutritional boost. After that, most goat-keepers rear kids by bottle-feeding them either with their mother's milk or with ewe-milk powder such as Vitamealo. (Don't use calf milk powder or fresh cow's milk – the fat content is too low.) This leaves the bulk of your goat's milk for you to milk off. (see *Milking*)

Gobies (see *Sea Creatures*) ***
These are active little rock-pool fish with a liking for an empty shell to live in, which should be supplied in their marine tank. There are several species, all suitable for the seawater aquarist.

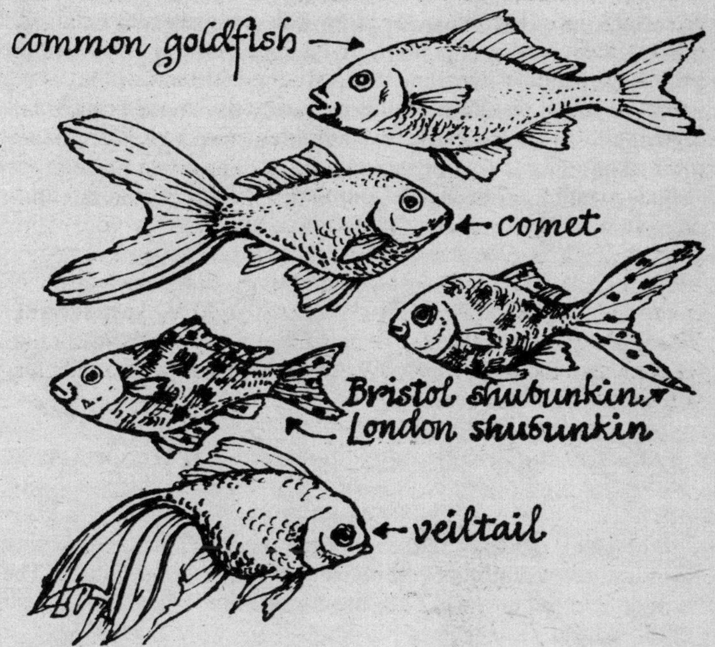

Goldfish (see *Fish*) ****
Hundreds of varieties of the **common goldfish** have been bred but generally the "fancy" variations are shorter-lived and less hardy than the original breed evolved a thousand years ago in China. Some fish have visible scales and others are smooth-textured, known as "calico".

The **comet** is the most robust of the fancy varieties, living for about fourteen years and thriving in outdoor pools. It is shaped like the common goldfish but has a tail as long again as its body, deeply forked at the end.

The **Bristol shubunkin** is shaped like the comet but is gloriously coloured in patches of red, black, violet, yellow and brown on a blue ground. It can be kept outside in the summer but should be brought in to an indoor tank for the winter months. Its cousin the **London shubunkin** is hardier, with the same splendid colouring but the shape of the common goldfish. It can safely be left out all the year round.

The **veiltail** is round-bodied with hugely developed fins and tail hanging like a veil. Easily damaged, it must be kept in an indoor tank. The same is true of the **fantail**, another round-bodied type with a deeply-forked tail, though the scaled variety is hardier than the calico.

The **black moor**, or **telescopic-eyed moor** is black, with bulgy eyes and the veiltail shape. Much too delicate to keep outside. The **oranda** and the **lionhead** both have weird mane-like development over the head.

Many owners consider the common goldfish to be more beautiful than any of its fancy cousins, for it has a simple, classic shape and its red-gold colour is a pleasure to see. Some specimens are yellow-gold and others patched with silver or black. They can grow to 20cm (8″) long in five years and can eventually double this length, by which time they are too big for tank life and need a garden pool. They can live for about 25 years.

Useful bottom-feeders for the garden pool are the **garden rudd** with flashy bright red fins, and the **tench.**

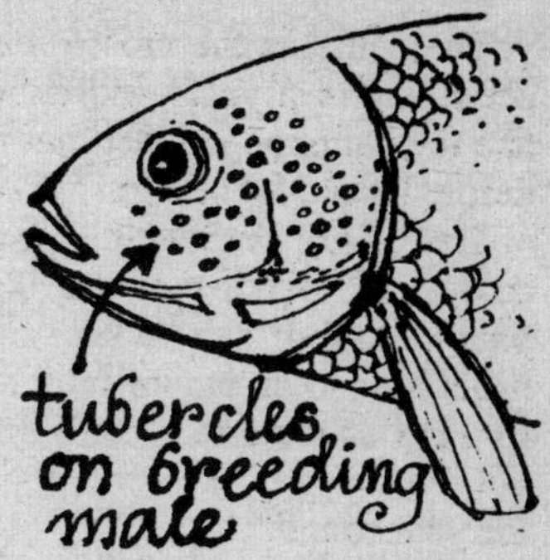

Goldfish can breed from under a year old but are best kept until they are 2½–3 years. Their spawning season is from April to September and the extravagant number of eggs laid allows for huge losses. It is impossible to sex goldfish until they are in breeding condition, when the whitish spots (tubercles) on the male's face show clearly. A pregnant female can be seen when viewed from above to be bulging with eggs (the "hard roe"). These are laid in thousands, and stick to water plants where they are fertilised by the male with the "milt" secreted in his "soft roe". Young goldfish are dull coloured, only attaining their bright gold when fully adult.

See *Fish* for general care.

Grasshoppers (see *Live Food*) **

These need a big cage or well-ventilated vivarium with plenty of room for jumping. Cover the cage bottom with some earth, leaves and clumps of grass, and put in some pot plants or branches. The atmosphere should be moist but avoid mould. Feed grasshoppers on lettuce, clover leaves, crushed oats and bran. They lay eggs in late summer and autumn. Their cheerful chirping noise is made by rasping the serrated edges of their long hind legs.

GRASS SNAKES (see *Snakes*) **

GREENFLY (see *Aphids, Live Food*)

GREEN (JERSEY) LIZARD (see *Lizards*) ***

Guinea Fowls (see *Poultry*) ***

Guinea fowls on free range are a great success from their own point of view but they reduce their owners to gibbering wrecks. It is something to do with the endless noise, the total devastation of the vegetable garden and the utter impossibility of catching the culprits. Grab one by the tail – and the tail comes out, evoking a nightmare about a treeful of bald guinea fowls, still uncaught.

Extremely hardy, guinea fowls make good aviary birds as their grey or lavender feathers with mad white polka dots all over them are very decorative. They also have red helmet-shaped knobs on their heads, and ear flaps. They lay small, hard eggs, sharply pointed at one end, which hatch in 24 days, the chicks being striped in black and brown. The eggs hatch very readily in an incubator but the resulting chicks must be kept strictly under control as they are quite wild and will run off in all directions if let out. The hen birds only lay after mating, so you must keep a cock bird with them if you want eggs.

The guinea fowl call is shatteringly loud, and incessant. As table birds they are delicious, somewhere between duck and pheasant in flavour. It seems odd that no shooting estate has yet established the guinea fowl as a game bird, for they fly as well as pheasants, breed as easily, weigh heavier and taste better. Their mixture of cheek and hysteria makes them ghastly pets, I think, but some people seem to like them. They eat absolutely anything.

how to hold a guinea pig

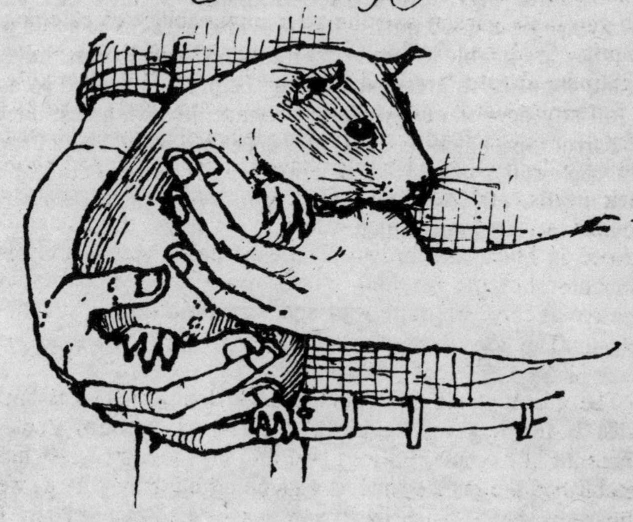

Guinea Pigs (Cavies) ****

These have been domesticated for hundreds of years and yet are still nervous animals, needing to be handled very gently and steadily. They must be picked up by shoulders and rump, not under their chests, which may damage their lungs. Although given to loud conversational squeaking themselves, they are alarmed by loud noises.

There are three basic breeds, the **English** (short-haired), the **Peruvian** (long-haired, needs a lot of grooming) and the **Abyssinian** (hair growing all ways in rosettes). Within these types there is a vast range of colours and the breeding of "cavies" can be a highly specialised business as they are widely exhibited. Showquality stock is advertised in *Fur and Feather* but animals purely as pets can be bought from dealers or shops more cheaply. Always look for clean, glossy coats and bright eyes, and buy young stock, as guinea pigs only live for 2½–3 years.

Housing should be as roomy as possible as young guinea pigs need ample space for exercise. The hutch described in *Rabbits* will do well for a guinea pig sow and her litter.

Guinea pigs need a lot of roughage in their diet and must always have good quality hay provided in racks to avoid wastage. They are unable to synthesise vitamin C and so must have a daily supply of fresh greenstuff, picked before it is needed so as to dry off any dew or melted frost. Cabbage, lettuce, grass, dandelions, carrot tops, chickweed and apple peelings are all liked. Proprietary pellets can also be used, or a mixture of crushed oats and bran, or a mash of cooked potatoes and bran. Water must always be available in a drinking bottle.

Newborn guinea pigs, unusually, are well furred and have their eyes open. They suckle from their mother but are able to eat solid food at two days old and can be weaned at 3–4 weeks, preferably by removing the mother to another hutch. The good development of the young is due to their long gestation period of 60–74 days (average 65). Litter size is usually two or three, though as many as seven has been known.

Female guinea pigs come into season every two weeks for 10–24 hours, and can always be mated within 48 hours of giving birth. They can breed as early as 4–5 weeks but it is better to mate them at 3–4 months. They are then not quite full grown but it is best for them to have their first litter before the pelvic bones fuse, thus avoiding difficulties in giving birth. The boar matures more slowly and should not be used for mating until 4–6 months old. To sex guinea pigs, see the diagram in *Rabbits*.

Guinea pig hutches should be provided with litter and straw in the same way as for rabbits (see *Rabbits* for details). The siting of hutches is important, for guinea pigs can suffer from heat stroke if kept in direct sunlight. They are also very susceptible to car fumes and their hutches should never be kept in a garage.

The laboratory use of guinea pigs has proved them to be very different from other rodents in their reaction to drugs. Many of the commonly used antibiotics are fatal to them, which makes them difficult to treat if they are ill. Some of the diseases they can suffer from are transmissible to humans, so always consult your vet early if your guinea pig seems unwell.

Hamsters ****

There are several species but only the **golden** is suitable as a pet. Hamsters hate each other and start to fight at a few weeks old, so must be kept alone.

Mouse cages are too small for hamsters. The minimum size should be 60×30×26cm high (24×12×10″) but should contain a private nest box supplied with hay. The hamster will also need wood to gnaw, for, like all rodents, the teeth grow continually and must be kept short through use. So, provide a small log and things to explore such as a cardboard roll or a smoothly opened tin. A jamjar on its side placed in the hamster's "dirty corner" will soon be used as a lavatory and can be cleaned daily, but the whole cage needs cleaning at least once a week, for hamsters store food and this must not be allowed to decompose. Litter the cage floor amply with peat or sawdust and provide paper to tear up for nesting. Some hamsters prefer this to hay.

Hamsters like a mixed diet of grain (but not whole oats, which may prick the inside of the cheek pouches), green vegetables and roots, dog biscuits, rabbit pellets, bread, cake and fruit, but they carry about more than they actually eat. Milk is also liked but must be given in a spill-proof dish. Clean water is best supplied by drinking bottle.

The female has a rounded tail end as opposed to the male's elongated one. At six weeks she will come into season every fourth day and mating may be tried at twelve weeks or more. Take the female to the male and watch her carefully. If in season she will "freeze" with her tail in the air and allow him to mount. If not, she will attack him, in which case, remove the poor chap at once before he goes off the whole idea. A successful mating will produce a litter of 4–7 bald, blind babies about sixteen days later. Unusually, they have teeth at birth. They grow quickly and should be taken from their mother at four weeks, or she may attack them. Separate them from each other by 6–8 weeks or they may attack each other. They can also breed at six weeks, so males and females must be segregated.

Hamsters are nocturnal, so suit people who are out all day as they are awake in the evenings. Never leave them on a table unattended as they fall heavily and injure themselves. Should an exploring hamster go under the floorboards or into the bowels of the sofa, don't start tearing things apart – put his cage on the floor where he disappeared and go to bed. He is usually back home in the morning.

Hamsters hibernate in the wild, and very cold conditions may send your pet into a deep, death-like sleep. Don't wake him up. Provide lots of bedding, keep cool and wait. When he wakes, maybe weeks later, increase the temperature gradually and provide plenty of food, starting with a warm wholemeal bread and milk mash. Conversely, high temperatures can send hamsters into a kind of convulsion where they go stiff and look dead. If disturbed they wave their heads and tremble. Leave strictly alone and they will recover of their own accord.

There is a National Hamster Council and local hamster clubs organise shows. Show animals must be in superb condition, scrupulously clean and friendly. No hamster ever won that bit the judge.

Hares X

Unlike rabbits, hares are solitary creatures. They do not burrow and rely on the speed of their long legs to escape from predators. Only 2–3 young are born in each litter but their eyes are open and they are well furred. Every hare makes a "form" or shallow depression in the grass where he or she lies still for most of the day, waiting for evening when the hares go out to graze. Leverets are born in such a "form", each baby soon making its own little form beside the mother's. If you come across a baby leveret lying still in the grass, don't pick it up or stroke it, for when the mother returns she will smell your scent on it and may abandon it. These beautiful animals are essentially wild and must not be regarded as pets unless you come across a captive-bred strain. The so-called Belgian Hare is a large brown rabbit.

HAWKS (see *Birds of Prey*) X

Hedgehogs **

Splendid little animals to have in the garden, prickly but with coarse grey hair underneath, long snouts, bright eyes. Adult males are about 24cm ($9\frac{1}{2}''$) long, females a little smaller. They are mammals, hibernating in cold weather but waking on warm winter days. Of nocturnal habit, they rootle for slugs, snails and other garden pests but also like bread, milk, cornflakes, fish, tinned cat meat, liver, sausages and chocolate cake, to mention but a few things. In other words, they appreciate human company and can quite well be regarded as pets if kept in a roomy compound in the garden. It seems unfair not to keep a pair, for hedgehogs breed twice a year, producing 3–7 blind, soft-haired babies in a litter (the hairs stiffen into prickles at about three weeks). The gestation period is about five weeks and the male should be removed before the young are born, for he may kill them or harass the female so that she neglects them. Baby hedgehogs are eating solid food at seven weeks old, and can leave their mother by this time.

A hedgehog's only defence mechanism is to roll up into a prickly ball – no good at all against the motor car. Rescue any hedgehogs found wandering on the road, for they are great friends to the gardener. Their teeming fleas bite humans but do not live on them. Don't shower them with insecticides, or you may kill the hedgehog – and don't use DDT or metaldehyde slug killers in the garden if a hedgehog lives there. (see *Live Food*)

HENS (see *Bantams, Chickens, Eggs, Poultry*) ****

Hibernation

Animals which hibernate go to sleep when the weather turns cold (the word literally means "wintering") and remain deeply unconscious until the spring. The heartbeat and breathing slow down until the animal is barely alive, making the least possible demand on its physical reserves.

Pets which hibernate need special provision. The first essential is to make sure they eat enough during the summer and early autumn to build up good stores of body fat. Cold-blooded animals such as snakes and tortoises, whose body temperature is the same as that of the air surrounding them, may be too torpid in a cold summer to bother about eating. This can be overcome by rigging up an electric light bulb near the ground in a rain-proof cover, which will stimulate the creature into healthy gobbling.

As the days shorten, find a strong, mouse-proof box big enough to fit your pet and his bedding – hay or straw, dry leaves or (ideal for snakes) dry leaf mould. The box should have ventilation holes but be securely lidded. Let your pet doze off where he usually lives and make sure he is really deeply asleep before you pack him away. Don't crate him up at the first signs of drowsiness, for a spell of warm autumn weather may well perk him up again. Once asleep, your pet in his box should be stored in a cold but frost-proof place such as a garden shed – not indoors, or your winter heating will wake him up again. Since snakes are great escapers, it's as well to put a sleeping snake's box inside another container such as a tea chest with a lid on – just in case.

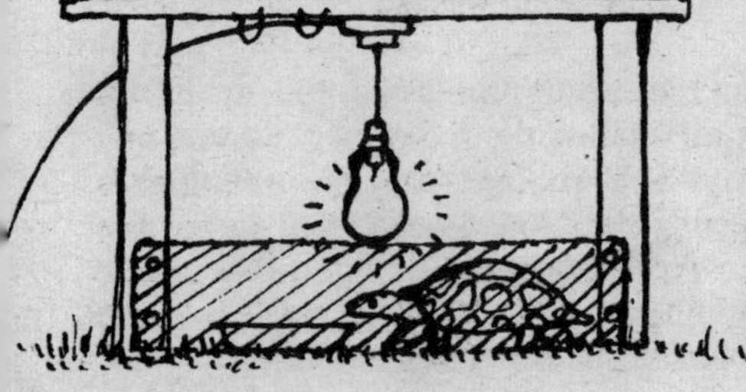

old table or purpose-made shelter. Sides can be fenced in. Use waterproof cable secured by large staples

Next spring, remember your sleeping pet and keep an eye on him. He will wake up slowly, looking rather hung-over. If his eyes are sticky, wash his face gently with some cotton wool and warm water – not soap.

Some mammals such as squirrels hibernate in a half-hearted way, catnapping through the coldest weather but waking up on mild days to forage for food, often previously stored. Such creatures must not of course be tucked away in boxes. They will curl up in their normal sleeping quarters and should be left undisturbed, though food and water should always be available.

Horned Toad (see *Lizards*) ***

This is not a toad, but a lizard. About 15cm (6″) long, it is broad, squat, very lumpy and ferocious-looking, covered with spines and with "horns" on its head, horrendously coloured black and yellow. Despite the fearsome appearance, it is a shy and harmless beast, liking to burrow and hide itself away, so needing a good 15cm (6″) of sand in a warm vivarium (see *Vivaria*). It eats caterpillars, grubs, ants, mealworms but not earthworms, and drinks dew from leaves. (see *Dew Dispenser*)

Horses and Ponies *****

As far as general care and handling are concerned, horses and ponies can be thought of as the same animal, although there are differences between them both in temperament and conformation, the most obvious being one of size. A pony can be of any size up to 14.2 h.h. and anything bigger is a horse. "H.h." means "hands high". A hand is 10cm (4″) and a horse is measured from the ground to his withers – where his neck meets his shoulders.

Pony breeds range from the tiny **Shetland** through **Dartmoor** and **Exmoor, New Forest, Welsh** and **Highland,** to name but a few. Many ponies are cross-bred and these are often very strong and hardy. Think carefully about what kind of pony you want before buying one. A pony for showing or eventing is quite different from (and more expensive than) a stocky, hardy pony for country hacking. The same is true of horses, which range from "fine" breeds to strong, weight-carrying types, culminating in the "heavy" horses once used for draught and agricultural work.

No pony or horse can be kept in a field with no further attention. Although grass is a natural diet, wild ponies exist on thin, mountainous forage where it is impossible for them to overeat. Lush meadow grass can easily produce laminitis, a painful and disabling foot inflammation, and working animals soon become over-fat and sluggish. It is, in fact, essential to have a stable, no matter how simple, for horses need to be very carefully managed.

Basically, a horse lives either "in" or "out", though some are turned out to graze during the day and brought in to sleep in a stable at night. A stabled pony is clipped and rugged so that he can work without getting too hot, his rug replacing his natural coat while resting. Ponies and horses living outside are not clipped but in cold weather may need the additional protection of a waterproof New Zealand rug. Stabled ponies must be mucked out every day, the droppings and wet straw removed and a clean bed put down.

The food a pony or horse needs will vary immensely according to his size and how much work he is doing. In spring and summer grass can be enough on its own, but when the growth slows down hay is essential, and must be fed throughout the winter. "Hard feed" such as oats, bran, proprietary nuts, barley and well-soaked sugar beet pulp must be provided throughout the winter and as a summer supplement for horses and ponies working hard. The quantities fed are impossible to detail as each individual animal is different. Fresh, clean water is essential, changed twice a day or supplied in a self-filling tank. Give water before feeding, none for an hour afterwards, and never allow a hot, thirsty pony to fill himself with cold water, or he will go down with colic, which is a violent pain in the tummy. You can suspect this if the pony is sweating, kicking at his stomach, pawing the ground and trying to roll. It must be treated at once with a "colic drench". Call your vet and meanwhile walk the pony about constantly. Never let him roll or he may twist a gut, which is fatal.

Stables, field shelters, fending and gates must all be strong and free of weak points that can be fiddled with until they give way. Avoid anything sharp or breakable, for horses are absurdly ready to hurt themselves, given half a chance.

Training a horse or pony is such a vast subject that it is outside the scope of this book, but there are a few common-sense points to watch. To start with, a young, unbroken pony will be much cheaper than a well-trained animal and it is often tempting to buy such a creature for relatively few pounds at a pony sale. Don't. Unless the owner knows what he or she is doing, the results can be disastrous. Horse diseases and malformations are legion, and it takes an experienced horseman or a vet to spot them, so always take an expert with you to look at a prospective purchase. If someone has already "had a go" at training the animal it may have been mishandled or got into bad habits which will be difficult to eradicate. The beginning of training is the building up of confidence between the pony and its owner, based on firm but gentle handling and, above all, on steady, kind, consistent behaviour by the owner. If the pony trusts you it should be easy to lead him by his head collar (especially if he has been used to a foal slip from early days), to move him round his stall or stable, pick his feet up for cleaning and catch him if he is out in his field. Without this trust, your task is heart-breakingly difficult, and that is why it is best to avoid half-trained young stock. For a beginner owner get a "safe" pony from a reputable source where inspection is welcomed. *Horse and Hound* is full of advertisements and local papers in country districts often have a "Horse and Rider" column.

A filly can be bred long before her training is completed. Study her characteristics carefully and select a stallion which is free of any faults she may have. Stud fees can be expensive but a good foal is far more valuable than an indifferent one. The gestation period is eleven months and in the summer it is probably better to let the mare foal down outside. In bad weather she should have a warm, dry foaling box with plenty of absolutely clean straw. Many mares resent any interference and they have a reputation for being able to "hold" a foal until they feel that circumstances are just right for its birth.

A horse or pony is possibly the most satisfactory of all pet animals, for it makes a whole new way of life possible. Riding is a subtle art involving constant learning, and the many events associated with ponies and horses are great fun – but the hard work of routine care can never be shirked.

House Martins (see *Wild Birds*) *

These are not pets, but it's a great pleasure if these summer visitors build their mud-dollop nest outside your bedroom window where you can hear the babies squeaking and see the parent birds flash past on their way up to the nest with beakfuls of flying insects. They lack the long forked tail of the swallow but make up for it with pretty feet, covered with white down.

Hummingbirds (see *Wild Birds*) X

Although very beautiful, these tiny birds are too difficult to be household pets, needing a high temperature and constant feeding from nectar dispensers.

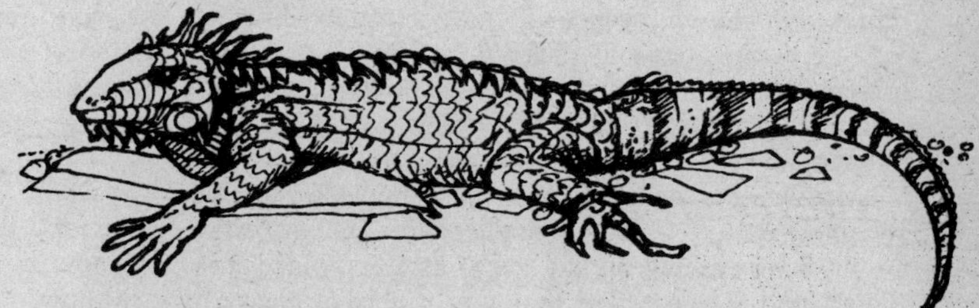

Iguanas (see *Lizards*) ***

The common iguana is an intriguing beast because it is exactly like a small dinosaur. Young ones are lizard-sized but beware – they grow to well over a metre (5′) long and will need a very large vivarium. Iguanas need light and warmth, so their vivarium should be kept near a window and should have a light bulb above it on dull days. They must have a large bowl of water and strong branches (see *Vivaria*). The temperature must not drop below 20°C (70°F).

Iguanas are largely vegetarian, eating lettuce, cabbage, banana, grapes, green beans and other soft fruit and vegetables, but will also appreciate a few mealworms or insects. A few spots of halibut liver oil can be added to their food about once a week and it is all to the good if they will eat a little finely minced liver or raw meat. A fine spray of warm water each day will make sure they do not get too dry.

Although iguanas seldom breed in captivity, a newly imported female may well lay eggs. These can be incubated in very slightly damp sphagnum moss and decaying wood at a temperature of 28°C (80°F) for about twelve weeks – if mum doesn't mind them in the airing cupboard, that is.

Some iguanas become quite tame and friendly, others never do. If you have a nervous one, make sure it always has food available, keep it safe from your noisier friends and don't handle it. Like most lizards, iguanas need to keep calm.

INSECTS

This is a surprisingly neglected group of creatures, many of which make beautiful and interesting pets. The life cycle is fascinating to watch as eggs hatch into larvae and then pupate, disappearing into the anonymous little barrel known in butterflies as a chrysalis. The emergence of the completely different, wet, soft adult creature poetically called a "nymph" is one of nature's most astonishing feats, and the immense variety of species is amazing, ranging as it does from barely visible flies to the huge, dramatic silkmoths.

Most insects are quiet, cheap, and simple to keep. They are surprisingly easy to obtain – suppliers are listed in *Useful Addresses* – but before taking on any insect as a pet, make sure you know what food it needs and that you will have plenty of it. Many species only eat one thing, and die in a very short time if deprived of it.

JACKDAWS (see *Crows*) *

Java Sparrows (see *Cage Birds*) ****

Neat, pretty little birds, these are cheap to buy, though the rarer colour variations can be more expensive, and easy to keep. Close-ringed birds have been captive bred and will be tamer and more ready to breed in captivity than captured wild ones. They need soft hay for nest-building and like rice in addition to the normal seed-eaters' diet.

Jellyfish (see *Sea Creatures*) **

These are best caught in deep water using a washing-up bowl so as to scoop the jellyfish up undamaged with plenty of water. "Beached" specimens are seldom any good. They will only thrive in a big marine tank with plenty of plankton.

KITTENS (see *Cats*. Also the name for baby rabbits) *****

Lacewings **

These are the transparent-winged, pale green insects which often come indoors on summer evenings, attracted by the light. They hibernate indoors and are quite easy to keep in a large box or muslin-topped aquarium. Both adults and larvae eat great quantities of aphids, so should be supplied with aphid-infested twigs or plants. They will also feed on sugar solution, sucking it from a wad of soaked cotton wool, and breed readily, laying curious little eggs on stalks.

Ladybirds **

These can be kept in a big jar but perhaps better in a container admitting more fresh air, such as a small tank with a muslin-covered top. There are many different species. A pair will mate throughout the summer but May and June see most breeding. Eggs are spindle-shaped, laid in batches of 3–50, usually under leaves but often in unsuitable places. Yellow-orange at first, the eggs change to greeny-grey just before hatching in 3–9 days.

The larvae change skins four times and pupate in 24 days. Ladybirds and their larvae are cannibals, cheerfully eating each other, the eggs and their shells, but a population explosion would doubtless occur were this prevented. Cannibalism is probably triggered by a shortage of aphids and acts as a self-regulator to the species. One observer reported up to 45 aphids eaten in a single day by one larva.

LAMBS (see *Sheep*) X

Leaf Insects ***

The best known of these is the **Javanese leaf insect,** a wide, flat, pale green creature of amazingly leaf-like appearance, even to the leafy legs. The male is about 6cm ($2\frac{1}{2}''$) long, the female nearly half as big again. They need to be kept warm, in a temperature of at least 20°C (70°F). They eat bramble and drink water from wet leaves, and will soon become tame although they may shed a leg if picked up carelessly. Leaf insects breed readily but with a long incubation period. For general care, treat as stick insects.

LEMMINGS (see *Steppe Lemmings*) ****

LEVERETS (see *Hares*) X

Limpets (see *Sea Creatures*) **

These die of hunger in a clean marine tank, as they browse on green algae. Owners of green tanks should collect limpets by chipping off the piece of rock occupied by the limpet. Any attempt to knock the limpet off the rock will result in a squashy mess.

LIVE FOOD

Big creatures eat little ones. Not being equipped with captive bolt pistols and frying pans, they eat them alive. This is not a disgusting idea, for the death involved is usually a rapid and expert one. A cricket disappears into a lizard's mouth with the speed of light and a snake can kill a rat far quicker than the reptile house keeper can. However, many people find this a difficult thing to come to grips with, and it is certainly something to decide about before you commit yourself to keeping anything which likes its dinner on the hoof.

The most common and useful live foods are described with rearing details, each under its own heading, but the following is a list of easily-found food animals.

From the garden you can get aphids (greenfly etc.), ants and "ants' eggs" which are actually pupae, slugs, snails, earthworms, spiders, beetles, caterpillars and moths. Many of these are worth keeping in their own right, and these, too, are entered separately. Nearly all of them eat smaller creatures. Be careful that no food insects are contaminated with DDT or other insecticides, for the poison which kills the insect will also kill the creature which eats that insect. A small dose may not be lethal at once but accumulated amounts collect in the animal's body fat, particularly in the case of hibernating insect-eaters such as hedgehogs, and have their deadly effect in the winter, when the animal is living on its reserves.

From ponds or streams there is water itself, teeming with microscopic life (see *Protozoa*). Daphnia or water fleas can be caught, cultured or bought, and from there onwards there is an endless chain of eat-and-be-eaten life. Newt larvae and young fish in all stages are food for bigger animals, though sticklebacks are to be avoided, as their spines may stick in an unsuspecting throat. Do not, of course, take bucketsful of tadpoles, for the frog is not as common as it once was, and needs cherishing.

For more domesticated supplies of live food, the mealworm is hard to beat, being digestible, non-smelly and easy to breed. Maggots are also a ready source of food but the bluebottle and house fly tend to favour smelly places to lay their eggs. Also, maggots are rather thick-skinned and sometimes "go right through" a small bird's digestion.

Drosophila (fruit flies) are an essential diet item for young reptiles, followed by small "hoppers" of locusts or crickets. Many schools breed these and some specialist suppliers are listed in *Useful Addresses*.

Unless you own a snake it is not necessary to regard mice and rats as live food, but their prolific breeding habits make them quite suitable for this purpose. Such a procedure will produce horror among friends and neighbours, for humans are no longer very well in tune with the natural way of things.

Lizards (see individual headings for *Crested Anolis, Geckos, Horned Toad, Skinks, Slowworms*) usually . . . . ***
There are over 3,000 species of lizard, quite a few of which lend themselves to keeping as pets. Their size ranges from a few centimetres to over a metre but they are nearly all carnivorous, eating a variety of insects, slugs, worms, grasshoppers etc. They are all strictly vivarium animals (see *Vivaria*) and slough skins several times a year. (see *Sloughing*)

Although unsuited to carrying about, a lizard is a fascinating pet because he and his vivarium form a kind of mini-world of a kind which existed millions of years before Man had evolved. According to where they come from, some lizards like to live in a higher temperature than others and it is important to get this right. All lizards must be provided with water, though their needs vary from droplets on leaves (see *Dew Dispenser*) to a small swimming bath.

The **common lizard** is native to Britain, found in dry places. The adult male is about 15cm (6″) long, light or dark brown with broken lines of darker spots, underparts of red or orange, and black spotted. The female is a little larger, with paler underparts. These lizards swim well, so must have access to ample water. They eat moths, spiders and insects. A pair will mate and breed readily, eggs being retained in the female's body until hatched, so 6–12 young lizards are delivered alive, about 3cm (1″) long. The parents ignore their babies so divine providence (you, in other words) must supply them with fruit flies until they are big enough to share their parents' food. (see *Live Food*)

The **Green** or **Jersey lizard** comes from the island of Jersey and is commonly sold in pet shops. It is green, very pretty, and the male has a blue throat. Beware – he grows to be 30cm (1′) long and must be kept in a big vivarium where he has room to run about. These lizards eat grasshoppers, slugs and worms. They lay 8–10 eggs in summer, which hatch in one month.

Sand lizards are very rare and endangered, though they were once common in Britain, so it is important not to take one should you be lucky enough to see it in the wild. A sand lizard is much bigger than the widespread common lizard, being up to 23cm (9″) long, but is similarly coloured, brownish, with broken stripes of black dots along the sides. The male can have a greenish tinge, and has yellow underparts, black-spotted. The female has a paler, cream-coloured tummy. Mating happens in May or June and from 5–12 eggs are laid in July, covered with sand or leaves and left for the sun to hatch. This happens in about a month, and the young ones fend for themselves. Should you acquire and breed a captive strain of sand lizards, don't be tempted to release surplus young ones willy-nilly into the wild, hoping to increase the population. Sand lizards only live happily in old-established sandy places with plenty of leaf mould and heather, often in areas where there are abandoned rabbit burrows. Seaside sand dunes do not form a really suitable habitat and ordinary gardens or parks are no good at all.

Zonures or **armoured lizards** are splendidly ugly and ferocious-looking, covered with sharp horny spikes, but intelligent as lizards go, and very tame, with no objection to being handled. There are several species but the one most commonly kept is the **armadillo girdled zonure,** 40cm (15″) long, greeny-brown. He is an obliging eater, enjoying meat scraps and bananas as well as the more usual lizard diet of insects.

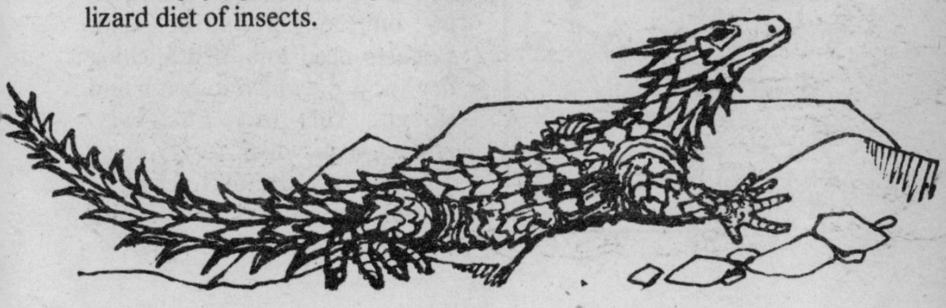

Some species of lizard regard their young as little meals. In these cases, have another vivarium ready and warm for the babies. If they arrive unexpectedly, a big glass jar in a warm place will make a good makeshift home for them – but don't let temporary arrangements become permanent, or your pets will die.

Never pick any kind of lizard up by its tail. They nearly all have a quick-release mechanism which leaves you holding a wriggling tail and the lizard looking like a long frog. The tail will eventually grow again but the new one always looks a bit stumpy and graceless.

Lobsters (see *Sea Creatures*) **

Be careful. These have claws like power-assisted secateurs. Feed them twice a week with strips of white fish and watch for the interesting process of moulting the shell (see *Sloughing*) which begins with a split down the back as if the whole beast is falling apart.

Locusts (see *Live Food*) **

These are quite interesting to breed and useful as a food for larger animals. The **African migratory locust** is the best sort to keep, as the **desert locust** is more susceptible to disease and has a longer life cycle, which may keep your hungry iguana waiting.

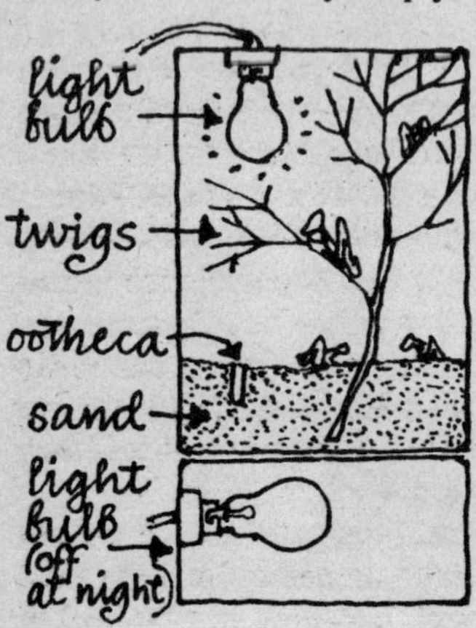

Locusts must be kept warm, ideally at 34°C (95°F) by day, dropping to 28°C (80°F) at night. This can be arranged by having two light bulbs, one in the locusts' cage and the other in a box underneath it, switching off the top one at night. Locusts need lots of branching dry twigs on which to perch and slough skins (see *Sloughing)* and sand in which to lay eggs. These are deposited in a capsule, or ootheca, sealed with froth. The new-hatched nymphs push their way through this and come to the surface of the sand after an incubation of sixteen days at 28°C, or as little as eleven days at 32°C. The nymphs moult five times before emerging as adults.

Locusts eat bran and fresh grass – but this must not be wet. In winter, wheat grains can be sprouted on damp blotting paper to provide a substitute greenfood. Escaped locusts will not reduce your garden to a mini-Sahara – British nights, even in summer, are much too chilly for them to live long outside. Locusts, however, go through a "plague" phase when they are brightly coloured, excitable and bunched together. When this collective aggression leaves them, they change to quieter colours, drift apart from each other and live quite placid lives.

MACAWS (see *Cage Birds Parrots*) **

Maggots (see *Live Food*) *

Maggots are the larvae of the housefly or its bigger cousin the blowfly or "bluebottle". These pests will lay their eggs on any unguarded meat or fish, infuriating housewives but serving a useful purpose in nature by clearing up any rotting corpses. Both the flies and their larvae are a useful live food for birds, reptiles and larger insects, and a maggot hatchery is a cheap and reliable way of feeding such creatures. The least smelly basis for such an enterprise is a half-used tin of dog or cat food left outside on a sunny day. When flies have laid their groups of yellowish-white eggs on it, put the tin in a simple framework (a cut-away cardboard box will do) covered with muslin. This prevents further laying and will contain the new-hatched flies. Maggots emerge from the eggs, eat the tinned meat, grow and pupate, much helped at this stage if you fill the tin up with sawdust. The maggots creep into this to pupate, and a handful of sawdust and pupae can easily be put in a vivarium to feed the occupants with flies when the pupae hatch. Adult flies can be kept for breeding in winter if they are fed milk and water on soaked cotton wool.

For feeding small orphaned birds, maggots should be slit longways with a sharp blade, or the end nipped off and the flesh extruded, as their skins are very tough and may not be broken down by the bird's digestive processes.

Maggots put into a deep-freeze as all fishermen know, will "come back to life" when thawed out, though they are as brittle as icicles when frozen. This is a useful fact in ensuring a continuous supply, but do make sure to label the containers, or you may be unpopular.

MARES (see *Donkeys, Horses and Ponies*) *****

Maybugs (also called *cockchafers*) *

These are large, tubby beetles which zoom about in early summer buzzing like small aeroplanes. Their larvae live in the ground for three years before emerging as adult beetles. These have a short but happy life eating leaves in the sunshine – which does the tree no good. They then mate, lay eggs and die.

Mealworms (see *Live Food*) **

These are the larvae of a small black beetle called *tenebrio molitor*, commonly sold by fishing shops. They can easily be cultured in a non-rusting container such as a plastic bowl, using a mixture of bran, brown bread slices and chicken meal. Some people put layers of sacking in this mixture and most add some raw carrot, potato or apple to keep the mixture moist. Introduce your mealworms – about 200 will do – and cover the whole thing with a big cabbage leaf. Keep the container in a warm place as the optimum temperature is 30°C (85°F). If mealworms get very cold they will hibernate in the larval stage, stopping your production line until they pupate in the following spring. The life cycle is from 4–6 months. Old food should never be discarded as the newly laid eggs, which will ensure continuity of production, will be laid on this. A well-run "factory" should always contain eggs, mealworms, pupae and adult beetles.

Mice *****

Splendid pets, portable, pretty and a fascinating demonstration of how Mendel's laws (the rules that govern heredity) work. Mice come in a vast variety of colours, both in recognised breeds and through casual cross-breeding of different coloured parents. Ordinary pet mice can be bought from shops or pedigree stock ordered from specialist suppliers. (see *Useful Addresses*)

Plastic cages are easiest to keep clean as metal ones tend to rust and wooden ones get gnawed to pieces, often releasing the occupants. Like all rodents, mice need something to gnaw to keep their teeth from growing too long, but a chunk of softwood or small branch will do. Mice smell because of the acetamide in the male's urine, so if non-breeding pets are wanted, keep two females. But even so, clean out cages at least once a week, supplying fresh sawdust and hay for bedding quarters. Cage design is a matter of choice, but certainly mice enjoy ladders and galleries, for they are active little creatures. An exercise wheel, too, is appreciated. A small nesting box, even for non-breeders, is essential.

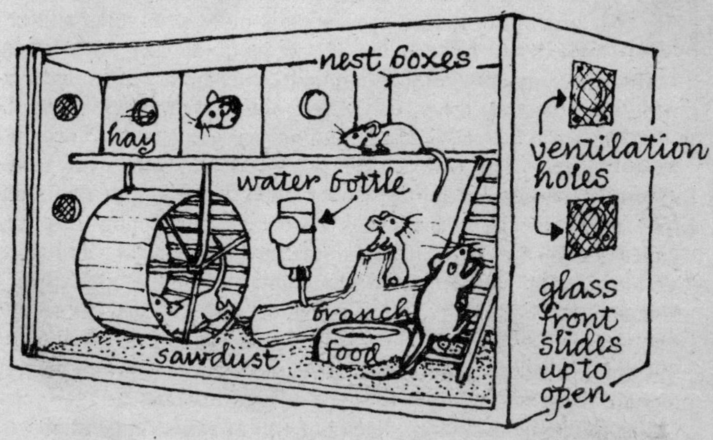

Mice eat whole grain – preferably oats – root vegetables such as swede, and like an occasional cabbage stalk. Too much greenstuff makes them smelly and "loose". Dried crumbled brown bread makes a good basic meal and a hard dog biscuit gives them something to chew and provides long-lasting food if you have to be away. Water should be available in a drinking bottle, not an open dish as they will scuffle sawdust into it. Get sawdust from a timber yard where no plastic laminates are handled, as powdered Formica etc. can be fatal.

Segregate young mice into males and females by six weeks old, as the female starts to come into season at about this time but should preferably not breed until she is ten weeks old. To sex mice, see the diagram under *Gerbils*. Female mice are in season every fourth or fifth night, when they will mate, producing a litter of 4–10 naked and blind babies between 19 and 21 days later. The babies should never be handled or even inspected as the mother may destroy them. It has been found that cannibalism of young may also be triggered by loud noises such as a road drill or vacuum cleaner. The young develop quickly and during this time the mother should be supplied with plentiful quantities of food. Many does also appreciate milk to drink. The young may be weaned at 21–28 days but by this time the mother will be about to give birth again, having mated immediately after the birth of the previous litter. For this reason it is a good idea to house her away from the male when it is first noticed that she is pregnant, only returning her when another litter is wanted. Never handle heavily pregnant does, for they can easily be injured.

Mice live amicably in colonies but will attack strange mice, so new stock should be put in an adjacent cage where they can be inspected but not touched until they have been accepted.

White mice are often surplus laboratory stock, bred sometimes with a deliberately induced drug resistance or a predisposition to a certain illness. Such animals should not really be sold as pets, for they will be impossible to treat should they fall ill. Coloured mice, on the other hand, are all produced for the "pet and fancy" trade, and are much safer.

Domesticated mice seldom live for more than two years, often only eighteen months.

Microworms (see *Live Food*) *

These are wonderful food for fish, even for quite young fry. These worms are too small to see individually but can be bred in small jars containing a spoonful of thin oatmeal porridge (cool, of course). Buy a starter culture from a fishing shop and put a drop in a porridge jar. In a couple of days the porridge will be gently heaving with invisible worms. Start a new jar for a continuous supply while feeding the existing worms to your fish. Do this by making a crisscross structure of headless matches, well soaked in cold water, on top of the porridge. The worms will climb on the matches, enabling you to feed your fish with worms and not with porridge.

Milking

Bring the cow or goat (or even sheep) into a weatherproof place or, if masochistic, take milking-stool, hay, feed and bucket out to her in the field. Tie her up if necessary. As a rough guide, she will need as much food as she can eat while you milk her dry, thus observing the principle of feeding a basic ration and adding a bonus for milk produced. Wash and dry her udder and sit beside her on her right-hand side. Goats are easier to reach if they stand on a bench, and some people prefer to milk them from behind, reaching between the hind legs.

To milk, nip the top of the teat between thumb and finger, trapping the milk contained. Close the remaining fingers from top to bottom. squeezing the milk down and out. Release the top of the teat to admit a new filling of milk and repeat ad infinitum (see diagram). It makes your hands ache for the first year or two but you get used to it.

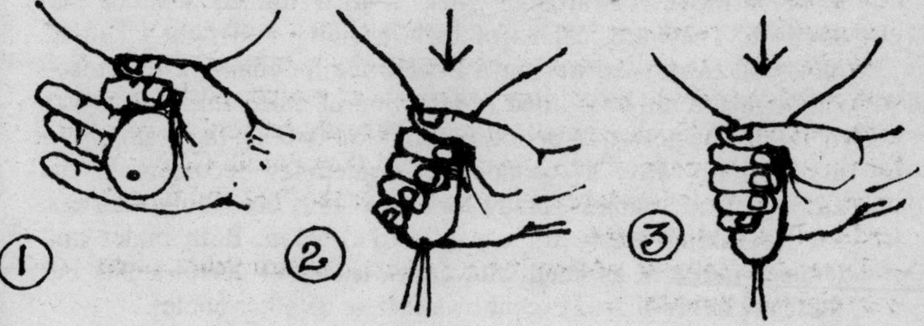

Strain the milk through a paper filter or two layers of butter muslin boiled after use, and cool it as soon as possible. Clots in the milk indicate mastitis, which you can often feel as a bubbling sensation in the teat as you milk, rather like air in a tap. Tell your vet. Blood specks in the milk are not uncommon in high-yielding cows on spring grass, but are not ominous.

Millipedes *

Unlike centipedes, these have two pairs of legs to each body segment and move about rather slowly, looking as if they are pouring themselves along. They have short antenna, are usually slate-grey and eat decaying leaves and other vegetation. Really rather dull.

Minnows (see *Fish, Live Food*) ***

Minnows can easily be caught in streams or ponds. Smaller than sticklebacks and spineless, they like to live in shoals and will not be happy in a tank unless kept in a group of at least half a dozen. They are very active, able to grab a quick mouthful of food from under the nose of a slower-moving fish.

Moles *

Nobody has managed to keep moles for very long in captivity but they are interesting to study for a short while and quickly become tame enough to feed from the hand.

Moles are almost, if not completely, sightless although they do have eyes. Their ears are invisible but acute and they sense movements in the soil very accurately. Their "hands" are spade-shaped and very strong and their dark grey fur has no "set", enabling them to move forwards or back in their tunnels without discomfort. They are jealous of their territory and fight all other moles of either sex, having a brief truce for mating, it seems – although the only recorded observation of moles copulating was in 1803. The gestation period is about four weeks, the young born in April or May, a second litter sometimes occurring in the autumn. Four babies usually form the litter, born quite hairless, the fur taking nearly three weeks to develop. Both males and females make nests lined with grass, leaves or cereal stalks but these are never shared except by a female with her babies.

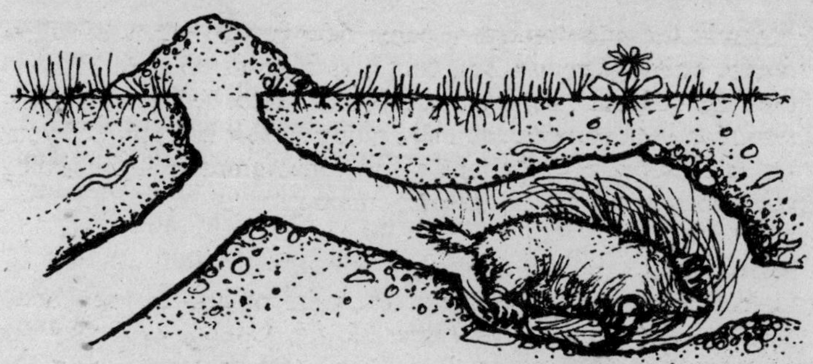

Gardeners hate moles for their habit of throwing up "hills" of tunnelled-out earth all over the lawn – a pity, for moles eat a lot of destructive cockchafer (maybug) larvae as well as earthworms and beetles. They are easily caught by digging and can be kept for a while in a big box or tank of earth, supplied with worms or mealworms (see *Live Food*), hay or dried leaves for bedding and water from an unspillable pot. Grasp them round the shoulders to pick them up, so they cannot turn their heads to bite.

Mongooses *

In the days of the British Raj, a mongoose was an essential pet in every household, kept to deal with the cobras much as a farmer will keep a good sharp Jack Russell terrier for the rats. My grandmother said her mongoose was a nuisance because he opened the lids of the inkwells and put inky paws marks everywhere. Not a common British pet.

Monkeys X

Although monkeys look appealing, they are very strong, dangerous animals whose sale is controlled by the Dangerous Wild Animals Act. They are unsuitable for keeping in ordinary houses because they are utterly undomesticated, impossible to board out for holidays, and apt to transmit nasty diseases to their owners. From the monkeys' point of view, house-dwelling robs them of freedom and the company of their own kind – and from the conservation angle, a monkey taken from the wild is lost to the species for ever, as "pet" life will not allow it to breed.

Moorhens (see *Wild Birds*) **

These neat little black water birds with red legs are apt to arrive uninvited on your duck pond and some poultry-keepers accuse them of drowning ducklings by pulling them under the water. I am not sure if this is true, but moorhens certainly can dive and stay submerged for some minutes, and it is funny to see their tiny black chicks bob up from nowhere. Although rather shy of humans, they help themselves to all usual poultry food.

MOTHS (see *Butterflies, Silkmoths*) ***

MULES (see *Donkeys*) ***

Muscovy Ducks (see *Ducks, Poultry*) *****

The most enchanting of feathered pets, the Muscovy duck has a sweet personality which takes a little time to discover. They are short-legged, rather cumbrous birds with black and white or grey and white plumage, red beaks and variously elaborate red wattles over their heads and faces. The drakes are big birds and can look like red-faced old colonels at their worst – and yet their appearance belies their gentle nature. Muscovies never quack, but keep up a conversational twittering among themselves, hissing in warning but greeting their owners with a croaky squeak.

Muscovies are kept like ordinary ducks but have the advantage of being excellent mothers, brooding and caring for ducklings most capably. They make obliging foster mothers for the young of other ducks.

If a Muscovy duck should be mated by a mallard drake (mallards tend to be quick workers), her eggs will hatch into huge hybrid birds which, though unrivalled as a table delicacy, will be "mules", of no discernible sex and unable to breed. This is true of any cross-mating between Muscovies and other ducks. Muscovy eggs can take longer than ordinary duck eggs to hatch (see *Eggs*), so be prepared to wait for at least five weeks.

Be careful when handling Muscovies. Although they are so friendly, they can panic if they are not used to being picked up, and they have very sharp claws which can easily cut your hand.

Mussels (see *Sea Creatures*) ***

A basic beast for the marine tank, the mussel is a good filter, cleaning up a cloudy tank beautifully but soon starving in a clean one. Best to return them to the sea after two weeks' duty unless they are wanted as food. Almost everything eats mussel flesh.

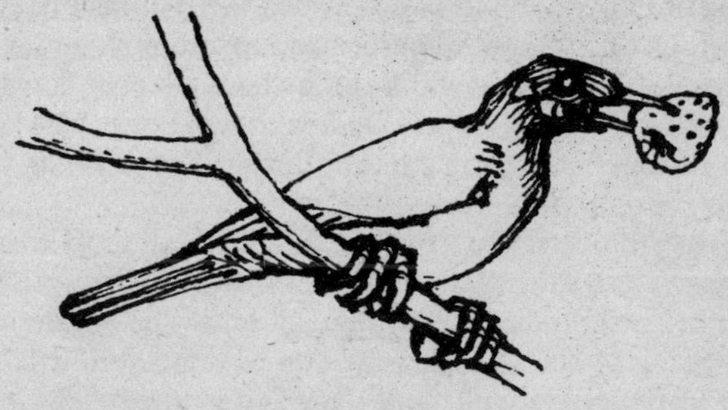

Mynah Birds (see *Cage Birds*) ****

Also called "hill mynahs" or "grackles", these are not seed-eaters like most cage birds and need different treatment. Glossy black in colour with orange bills and yellow head wattles, these are "soft-bills", which means they eat fruit rather than hard seeds. They also drink a lot, which makes them messy pets to keep, and the ordinary "prison bars" cage is no good, as the floor surrounding it will soon be awash with chucked-out food and water.

Mynahs are best kept in the largest possible boxtype cage with a wire front (see diagram). This should be at least a metre or yard across, and preferably bigger. Some people build in an alcove in a room to provide spacious and very attractive living quarters – and it is worth making a fuss about getting this right, for mynahs are expensive birds and can't be taken on lightly unless you've just won the pools.

It is best to buy a young bird between the months of June and September, as this follows the breeding season. A young bird will look rather "sooty" as the gloss on the feathers only comes later on, and he will probably open his beak wide when he sees you, gaping for food. Most people like to teach their mynahs to talk, but mature birds seldom learn.

Don't rush into letting a newly arrived bird out to fly freely in a living room. Keep him caged for at least two weeks so that he is secure in his surroundings, and then get him used to perching on your finger inside his cage. He may still be difficult to catch when let out, in which case wait until dark. Your night sight is better than his and you can pick him up easily. Mynahs are messy, wet droppers, unlike the desert-dwelling budgerigars, so there will be a lot of clearing up to do. Move any sharp, pickable-up objects such as drawing pins and draw the curtains so that your bird doesn't dash himself against the glass.

Mynah birds need to eat almost constantly, so food has to be available all the time. A proprietary meal will do as a basis but a variety of fruit such as apples, oranges, stoned plums and berries must be given. Lettuce, watercress and ripe tomato are also popular. Dates and figs can be given as they are, but sultanas, currants and raisins should be soaked first. Live food is not essential except for breeding birds, as the chicks live almost entirely on insects. Mealworms are very popular with all birds and crickets are a splendid food, using the newhatched nymphs to feed mynah chicks. (see *Live Food*)

Mynahs are almost impossible to sex, which is one reason why they are difficult to breed in domestic conditions. Aviary-kept birds will breed in a casual sort of way, incubating their two pale blue eggs intermittently and often letting them cool. Despite this the eggs will hatch in a total of about fourteen days of warmth, no matter how often interrupted, and the parents feed the blind and naked chicks with insects.

Newts ***

There are several species, some found commonly in Britain. It seems reasonable at present to catch wild specimens to keep as pets (see *Amphibians*) but don't be greedy and take too many. All else apart, newts are apt to be bullies and need plenty of space. A tank 60 × 30cm (2 × 1′) is only big enough for three newts at the most.

Newts live mostly in water and an aquarium tank for them must contain three or four water plants, preferably of the fern-leafed type, as newts lay their eggs singly on the edge of water plant leaves in spring, folding the leaf edges over protectively. These eggs hatch in 1–3 weeks into newt tadpoles, slim and transparent. They attach themselves to water plants by long sucker-ended tentacles from the sides of their jaws and stay put, developing front legs first then hind ones, and replacing their external gills with lungs. They keep their tails and are complete miniature newts in a few weeks, usually mature by the autumn. In winter all newts hibernate and it is unwise to try and keep them awake. Put the tank in a cool place but where it won't freeze solid.

Mature newts spend a lot of time out of the water so their tank must contain big stones that come above the water surface. Living in an outdoor pond, newts like to bask on water lily leaves.

Heat is the newt-keeper's great enemy, for newts die very quickly if the air or water temperature gets too high, so never keep their tank on a sunny windowsill. Don't handle them more than necessary, for the heat of your hand will cause them distress. Their skins are part of their breathing system, and are very sensitive.

Newts eat earthworms and small snails or bits of meat (but don't leave these in the water to rot). On land, they eat insects. They slough their skins regularly. (see *Sloughing*)

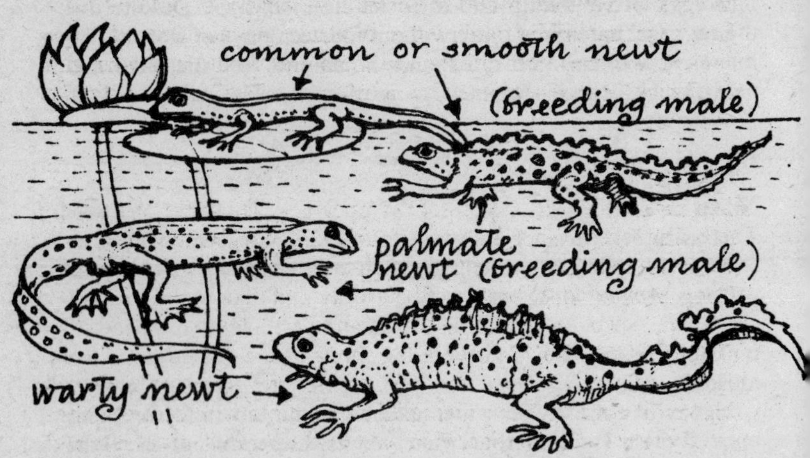

The British newts are the **common** or **smooth newt,** the **palmate newt** and the **crested** or **warty newt.** Because their colouring varies quite a lot these have sometimes been divided into separately-named species but it really seems that there are only these three.

The **common newt** is 8cm (3″) long, brown with black spots, orange or yellow underneath, also spotted. The male has a crest down his back in the breeding season. A shy animal, this spends most of its time under stones, hiding.

The **palmate newt,** also 8cm (3″) long, is prettier than the common sort, greeny-brown, less spotty, the male brightly coloured with a blue band on the tail and a handsome crest in spring. "Palmate" means "palmed", because of the "palm" of webbed skin between the toes of the male in spring. After breeding, when the animals normally take more to land living, this palm shrinks to a mere fringe along the toes.

The **crested** or **warty newt** is both crested and warty, a large, lumpy newt 15cm (about 6″) long, much the biggest of the British kinds. Dark brown, black-patched, orange underneath.

There are many foreign newts but the only one to breed readily in captivity is the **marbled newt,** a big European chap 20cm (8″) long which will bully the others and must be kept only with companions of his own sort. He comes in a variety of colours from grey to green, boldly patterned with black, the male crested. Less aquatic than the others, these are sometimes found in woods and heaths in France. Reputed to interbreed occasionally with the warty newt.

Ocelots X

The smallest of the big cats, the ocelot is about the size of an Alsatian dog, spotted and very graceful. It had a great vogue as a household pet some years ago but its sale is now restricted by the Dangerous Wild Animals Act.

Octopuses (see *Sea Creatures*) **

Don't collect big ones, for they climb out of tanks and will frighten Auntie to death if found in a knitting bag. Little ones eat small crabs and like a sheltered life, building a wall out of bits and pieces if no suitable rock exists.

Oiled Birds X

Not pets, but rewarding birds to care for if you have patience and don't mind the high risk of the bird dying. First decide whether the bird is (a) totally covered with oil; (b) two thirds covered; (c) one third covered.

(a) There is no hope of saving the bird and it should be humanely destroyed.

(b) If the bird is fat and healthy it may have a chance. A thin bird in poor condition is unlikely to survive. If your bird can't stand up there's not much hope.

(c) A good chance of survival. In this case get advice if you can from an Animal Rescue service (see *Useful Addresses*), a vet or the RSPCA. Don't attempt to clean it yet. The bird will have swallowed a lot of oil through trying to preen its feathers, and this is the chief hazard. Wrap it in an old woolly or something to stop it preening and keep it warm in a cardboard box lined with newspaper. Feed it strips of raw white fish, if necessary opening its beak and gently poking them in. When there is no longer any black oil in its droppings and it is feeding itself normally, you can wash it. Use a 1% solution of cheap washing-up liquid (Village, Keynote, Winfield or Co-op are good, posher ones less so) in hand-hot water, i.e. 150ml ($\frac{1}{4}$ pint) to 14 litres (3 gallons). Take the bird's restraining woolly off and immerse it up to its neck, one person holding, another washing. Swish the wings carefully through the water, using a toothbrush to scrub the head feathers. Part the gummed-up feathers to let water through. It will take two or three consecutive detergent baths to get the bird clean, and it must then be rinsed very thoroughly or its feathers will not be waterproof. A shower attachment is useful, spraying water "up" the feathers to the skin. When the bird is dry it should be put in a pen or shed with access to clean water so that it can bathe and preen. With good feeding and ample clean water it should be waterproof in about a week. It must then be released – not, of course, on an oil-polluted beach.

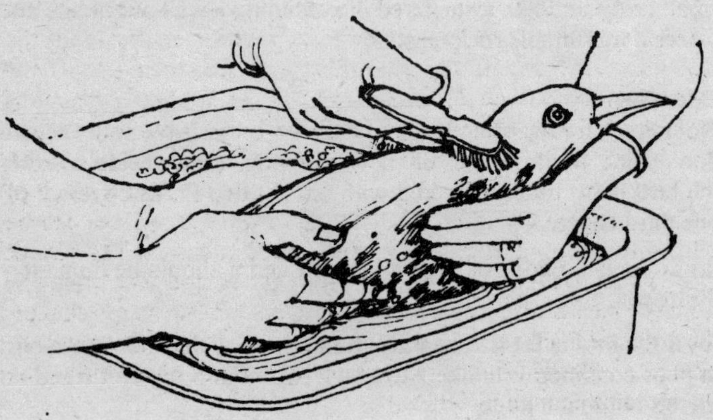

Orphans

Rearing an orphaned creature demands immense patience and hard work. The failure rate is high and in many cases it is kinder to take the small creature to your vet for painless destruction.

Baby birds found squawking on the ground are usually not orphans and are best left to the parent birds. If the parents are known to have been killed, efforts can be made to rear the young. Nestlings are of two basic types: those which are hatched bald and blind, totally dependent on the parent for food and warmth, and those which, like ducklings, can run after the parent. These, of course, are hatched able to see and well covered with down.

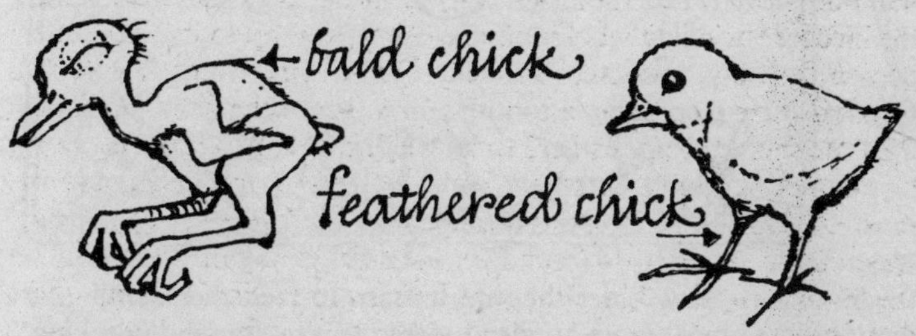

Bald nestlings should be kept in their nest if it can be salvaged, or in the best substitute you can make, the whole thing kept very warm, between 27°C and 32°C (80–90°F). The airing cupboard, if not efficiently lagged, may do, or use an infra-red bulb.

Very small birds will "gape" when anything comes near, but older ones may have to have their beaks prised open gently for the first two or three feeds. The vital thing is to feed them *every hour* from dawn to dusk or they will definitely die. Caught napping as one always is, the easiest food is sieved hard-boiled egg (which must be boiled for half an hour to make it digestible) mixed with four times the volume of crumbled sweet biscuit. Add a pinch of salt and moisten the mixture with a little water so that it can be rolled into pellets and fed with blunt-ended tweezers. The parent birds would feed the babies with insects, so the ideal thing is chopped mealworms, caterpillars, spiders or anything similar. Maggots are thick-skinned and sometimes "go right through", so should be squeezed out of their skins first. Don't give water – the diet contains enough moisture.

Feathered babies also need heat but can manage in a box with an overhead 60–watt light bulb. They will run about and can eat chopped hard-boiled egg and proprietary chick crumbs from a pet and garden shop. They must have water in a shallow dish – a lemon squeezer is ideal, the hump in the middle preventing them from falling in.

As chicks grow older they must learn to feed themselves and may need to have food dangled before them to stimulate pecking. Once feathered, wild bird chicks need a shallow dish of water. Legally, all wild birds must be released to the wild when recovered, but this is very difficult to do. A bird dependent on humans is not able to fend for itself and it may be argued that psychological unfitness is as crippling as physical inability.

Some birds – pigeons and gulls, for instance – regurgitate food for their young, so the best chance is to make a sloppy mixture of the above-mentioned ingredients and milk, and administer it by dropper or small spoon. For gulls, add some mashed raw fish and cod-liver oil. Baby birds of prey eat raw meat only. Cull chicks from a hatchery are ideal, or pieces of mouse, but they will take small strips of raw beef or chicken. Don't give water and don't try to feed them like seed-eaters or you will kill them. Get in touch with experts such as the Wild Life Rescue Service. (see *Useful Addresses*)

Orphaned animals, domestic or wild, need milk, for they are mammals, dependent on their mothers' milk. The young of such creatures as reptiles and amphibians, fish and insects are seldom "orphaned" since they are nearly all independent of maternal care at hatching. The problem of supplying the right kind of milk is critical, for cows' milk is meant for calves and is not necessarily acceptable to the young of other species.

Little is known about the milk of small mammals but some work has been done on the milk of cats and dogs, showing that it is much richer than that of cows. Findings vary but average figures suggest twice the fat content of Jersey cows' milk. Orphan puppies and kittens are unlikely to survive if they have not suckled from their mother at all, as colostrum is essential to ward off infections, but after that the best substitute is probably a commercial veterinary preparation such as Esbilac or Lactol, made up according to the manufacturer's instructions. Dried milk intended for human babies such as Ostermilk can be used but should be made up at twice the stated strength. Dire emergencies always happen when the shops are shut, so use the top off the milk with a little raw egg yolk added, warmed to blood heat by standing a container of the mixture in hot water. Never heat milk for young things in a saucepan over direct heat – it always upsets them. A doll's feeding bottle is ideal for kittens, or the rubber innards of a fountain pen. Your vet can supply you with a proper dropper – a bigger teat, of course, is needed for large-breed puppies.

A more precise formula devised by Pedigree Petfoods is as follows:

	Large quantity approx. 24g (¾ oz)	*Small quantity* approx. 6g (¼ oz)
Fresh cow's milk	600ml (1 pint)	1 teacup
Single cream	1 teacup	¼ teacup
Egg yolk	1 yolk	¼ yolk
Cod-liver oil	2–3 drops	1 drop
Sterilised bone meal	1 teaspoon	¼ teaspoon
Citric acid	½ teaspoon	1 pinch

The last two items can be omitted if not available, and you must *not* use gardener's or fertiliser bone meal. Ask a chemist for feeding quality sterilised bone meal. The babies will have to be fed day and night for a fortnight, after which things are a little easier. Don't take the job on unless someone will help you with night feeding – see table on next page.

Age in weeks	*Feed frequency*	*Grams of food per day (not per feed) according to animal's weight* Body weight in kilograms: .25 (½ lb)	.5 (1 lb)	1 (2 lb)	1.5 (3 lb)	2 (4 lb)	2.5 (5 lb)
1–4	2 hourly (3 hourly for strong puppies-	42	85	170	255	339	424
5–13	3 hourly (4 hourly for strong puppies)	56	113	226	311	453	538
14–24	3 hourly in day, one feed at night	70	141	283	424	566	707
25 on	3 hourly in day, no night feed	85	170	339	509	679	850

Introduce a little puppy meal and tinned meat when the animal is lapping from a saucer at 3-4 weeks, and gradually increase solid food, but keep bottle feeding for at least six weeks.

Goats' milk, with its high fat content and small fat globules, is particularly suitable for feeding smaller animals. Baby rabbits and "pippin" pigs do well on it, and some dog breeders keep a goat just for the milk to rear puppies. Use it undiluted, warmed to blood heat by standing in hot water. Goats' milk freezes well but must be thawed slowly.

Puppies and kittens need feeding six times a day for the first few weeks, gradually decreasing the frequency of feeds but increasing the amount given per feed. Solid food can be introduced as for normal animals.

For the first five days of life, puppies and kittens tend to be the same temperature as their surroundings, as if they were cold-blooded animals, so it is vital to keep them warm. Aim at a temperature of 30–32°C (85–90°F) at first, gradually dropping to 24°C (74°F) at four weeks.

The reason why puppies and kittens stay clean and dry in their nest is that their mother licks their bottoms carefully after feeding them. This licking stimulates the babies to urinate and defecate, and the waste matter produced is licked up by the mother. She has no dislike of doing this but will stop once the babies begin to eat meat. Orphaned puppies and kittens are unable to pass urine and faeces without the stimulation of licking or a similar sensation, so it is up to the owner to enable the little creatures to relieve themselves. No, you don't have to lick them! But after each feed you must stroke each one's bottom with a wad of slightly moistened cotton wool, which will mop up the resulting output. To make sure you don't miss one out, put the "done" ones into a separate box until the job is finished, when you can return them all to their nest. In about four weeks they will start to explore and will then begin to make puddles on their own, outside the nest – but it is as well to keep up the after-dinner mopping until they are eating solid food.

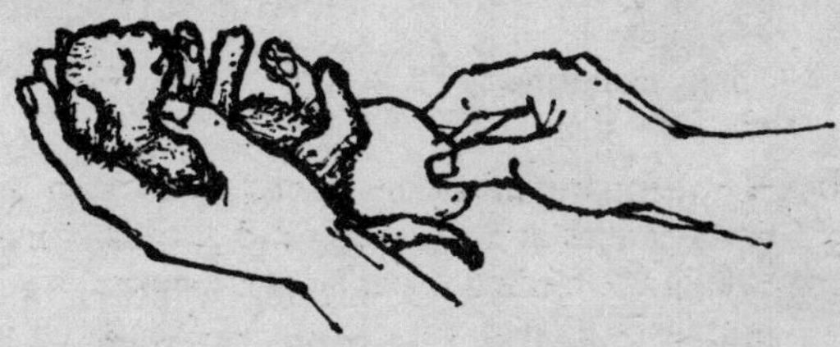

The National Foaling Bank and the Donkey Foaling Bank exist to help motherless or rejected foals, whether of horse, pony or donkey (see *Useful Addresses*). Donkey milk is low in fat and high in sugar, so can be imitated with Ostermilk and added glucose, feeding 300ml ($\frac{1}{2}$ pint) every three hours from 4 a.m. to 11 p.m., gradually decreasing the number of feeds but increasing the amount.

A hand-reared youngster will grow up to regard you as its mother, so the responsibility is great. Don't start unless you can keep it up, for the small thing's life is entirely in your hands.

Otters X

Although marvellous to look at, intelligent, charming and much popularised by *A Ring of Bright Water*, otters are not suitable animals as pets. They can be playful at one moment and bite you to the bone the next. If you like them, go and see them at zoos or the Otter Trust in Suffolk (see Useful Addresses). Everyone who works professionally with such animals is in desperate need of support from the public. Otters are classed as a legally protected species.

OWLS (see *Birds of Prey*) *

Parrots (see *Cage Birds*) *****

Although very expensive to buy, parrots are very intelligent, long-lived (over sixty years), beautiful and amusing. They can be kept at ordinary room temperatures in a large cage. It is important that a cage should be tall, for parrots climb about a lot. At least two perches should be supplied, one higher than the other so that the bird gets exercise and interest. A bored parrot soon gets out of sorts and often turns to feather-picking, which is a distressing habit. A rough branch with bark on it gives a parrot something more rewarding to chew at, and may also contain tasty insects.

The African grey parrot is the most reliable talker. **Macaws** and **cockatoos** tend to emit shrill screams and this makes them rather unsuitable as pets. **Cockatiels,** on the other hand, are charming little parrots, friendly to other species such as budgerigars and finches, and easy to keep. They will breed freely in a large enough aviary, rearing several broods each year, but don't expect much if they are kept in a small prison. Their only disadvantage is their high purchase cost.

A newly imported parrot may be rather delicate until it gets acclimatised, but once used to our weather, parrots are hardy birds. Never keep a cage in a draughty place, however, and always keep perches, food trays and drinkers very clean. Parrots enjoy a spray with warm water provided that there is no draught while they are drying.

A parrot seed mixture can be bought or made from one part white or brown millet, one part canary seed, one part sunflower seed and a quarter part each of oats, wheat, hemp and maize. Any kinds of nuts will be enjoyed though brazils and other very hard ones should be cracked. Macaws have big, strong beaks and can crack their own nuts, but find canary seed and millet too small to manage. Apple cores are popular for the pips as well as the fruit, as are plum and peach kernels. Give fruit, raw root vegetables and cabbage stalk, watercress or dandelion leaves, and don't forget a little coarse grit.

Parrots regurgitating food are going through a courtship ritual unless they look ill and huddled, drinking a lot but eating nothing. In this case keep the parrot very warm and call your vet.

Dried egg products included in some seed mixtures have occasionally proved to be infected, causing enteritis in parrots and other birds. It is best to avoid these mixtures.

Partridges **

Partridges can be reared in the same way as pheasants but, lacking the decorative qualities of the pheasant, they are less rewarding. And I have never seen one in a pear tree.

Peafowl (Peacocks) (see *Poultry*) ***

These are very beautiful, although it is only the male who has the spectacular tail. They are also very noisy, emitting a shrill scream which is likely to make you unpopular with the neighbours. They are expensive birds to buy but a pair will breed successfully in the right conditions. The hen bird starts to lay in her second summer although she and her mate will not attain their splendid adult colours until their third year. Chicks can be left with their mother, who must be protected from foxes if she incubates her eggs in the open. Alternatively, a large domestic broody hen can sit on six peafowl eggs. They hatch in 28–30 days.

Peafowl can be allowed free range if your vegetable garden is protected from them, or they can be kept in a large aviary, preferably with trees growing in it. Perches of 5–8cm (2–3″) diameter must be provided, high enough to keep the male's sweeping train off the ground.

A high protein diet is needed for peafowl, and turkey pellets are the easiest ready-made diet if they will accept them. If not, use mixed corn (wheat, maize and barley but not oats) with a little minced raw meat, some dried fruit and plenty of fresh greenstuff.

Free-range birds will find a lot of grubs and insects, and mealworms could be bred for aviary birds. (see *Live Food*)

Peregrine falcons *(see Birds of Prey)* X

Periwinkles *(see Sea Creatures)* **

These are the little shells found in millions on breakwaters and under piers, or on weedy rocks. They are hardy and live well in marine tanks but tend to climb out unless securely lidded.

Pheasants (see *Poultry*) ***

Although pheasants exist as a wild species, they are so widely bred by the shooting estates that it is doubtful whether the species is truly feral nowadays. Domesticated pheasants in their many species have been kept for so long that they are a well-established branch of poultry-keeping.

Most pheasant eggs hatch in 23–24 days, though some of the exotic species take 27–28. They are not easy to hatch in incubators as they are very susceptible to changes in temperature, and a good bantam makes a better job of it. Even so, the newly-hatched chicks are often very wild and need to be put in an escape-proof pen with their foster mother. Use a small-mesh netting, for pheasant chicks are very small. Feed them on hard-boiled egg to start with, going on to chick crumbs and finely chopped greenstuff.

Decorative species such as the **gold** and **silver pheasant** make beautiful aviary birds and will breed well, but their distinctive colouring would make them a prime target for a fox if let out, so these cannot be regarded as free-range poultry. Game pheasants can be hatched to release, but these must still be fed regularly if they are to survive.

Pigs X

One or two people have claimed that pigs, with their undoubtedly high intelligence, make good pets. "Pippin" piglets can be bottle fed with goat's milk and may grow up to be tame and affectionate – but the sheer size and devastating energy of pigs must keep them firmly in the farmyard.

Pigeons (also see *Doves*) *****

If racing pigeons are to be included in this heading, there is a vast subject involved. The breeding of racing pigeons is a highly specialised art, and the sport itself is extremely competitive, with big prize money at stake, and good birds fetching high prices. The Royal Pigeon Racing Association (see *Useful Addresses*) is a very helpful organisation and will encourage young pigeon-fanciers even if they don't know much about it. They are also anxious to hear news of any missing bird, so if you find a racing pigeon hurt, exhausted or even dead, note the number stamped on the underside of its wing and tell the Association, or return the ring from a dead bird's leg. A live bird will be paid for to cover its feed and the cost of returning it, so if you are able to help one of these long-distance travellers, rest assured that the owner will be more than grateful to you.

There is an almost endless range of pet pigeons apart from the racing kind, some of them so crazy in shape and feathering that they are hardly recognisable as birds. Many people find the extremes of specialized breeding distasteful, but there are several well-established varieties which are hardy and easy to keep. When people speak of doves they usually mean the white fantail pigeon which look so pretty kept in a white-painted dovecote. (see *Doves*)

As every gardener knows, pigeons eat almost anything, and a lot of it. Ready-made pellets are the simplest diet, supplemented with fresh greenstuff and clean water, but pigeons will eat any kind of cereal, whole or rolled, and any large seeds such as sunflower, maize or chick peas. Maize is a very useful food for breeding birds as it contains a lot of vitamin A, an essential element for fertility. Pellets contain some animal protein and pigeons on an all-grain diet can often be seen pecking at stale droppings because they contain vitamin B12, more usually found in meat or fish.

Pigeons lay two eggs which both parents incubate. New-hatched young are bald and hideous, fed by their parents on a highly nutritious regurgitated "pigeon milk".

Unless you grow vegetables, pigeons can be allowed to live freely out of doors, or they can be housed in an aviary. Racing pigeons live in a proper "loft".

Polecats (see *Ferrets*) X

The wild polecat was nearly wiped out by its chief predator, the gamekeeper, but kept a foothold in the hilly parts of Wales. It has been reintroduced in Scotland, where it is slowly spreading. Since it is so endangered, anyone finding an injured young polecat should get expert help as quickly as possible.

PONIES (see *Horses and Ponies*) *****

Poultry usually ****

Because hens, ducks and other poultry are useful, they are not always rated very highly as pets, but in fact they are friendly and companionable and, in many cases, very beautiful.

You can't keep poultry in a flat but even quite a small garden is

big enough for a few birds. Since all of them like greenstuff to eat, it's fatal to mix poultry and serious gardening, though they do no harm to grass provided they are not overcrowded. Some restraint, then, is usually necessary. This can be a caged-in run if space is short or strategically-placed wire fences to separate chickens from strawberry beds if they are allowed general free range. The fact is that poultry are immensely accommodating and will fit in with whatever arrangement you choose to make. The simple essentials are, first, open space for exercise and, second, shelter for the night. Some birds prefer to roost in trees but the less independent need a shed or hut which can be closed against foxes. This shed can be any size but must be supplied with bars for the birds to perch on and straw-lined boxes for them to lay eggs in.

Proper chicken huts have boxes along the side of the building with access from inside for the hens and lids on the outside for the owner to collect the eggs, but nothing so elaborate is needed. Hens will lay eggs from the age of about 21 weeks onwards, choosing any secluded, private-looking spot in which to make a nest. They love a stack of hay or straw bales, for instance – but the owner has to climb all over it to find the eggs.

Cereal forms the basis of all poultry diets but birds also need protein if they are to continue to lay eggs. Most people buy a ready-mixed layers' mash or pellets, but domestic birds, especially if they are on free range, can pick up a lot of good food in the form of live insects, worms and grubs, and household scraps such as bacon rinds and bread crusts are valuable. They won't eat raw potato peelings but cooked potatoes are gobbled up eagerly, specially if a few meat scraps are thrown in. The third part of a good poultry diet is greenstuff, and this is just as important as wheat or pellets. Left to themselves, birds eat quite a lot of grass and wild plants, and a regular supply of fresh greenstuff to penned birds pays off in terms of glossy feathers and beautiful eggs.

Clean water is, of course, essential at all times, for birds drink a lot, specially in hot weather. They make a terrible mess of an open dish of water, and tip it up by treading on the edge, so it is worth buying a proper poultry drinker (see diagram). This must be emptied and refilled twice a day. Don't just top it up – dirty water is a great spreader of disease.

Birds have a simple digestive system consisting of a "crop" or first stomach where the pecked-up food is stored, followed by the "gizzard", which is a kind of corn mill where the food is pulverised into a soft form ready to be assimilated by the gut. The crop is at the base of the neck and you can see it bulging in a well-fed bird, particularly in ducks. The gizzard only works when it is lined with little hard stones, and birds find these in the ground as they peck and scratch. Less obviously, birds also need "soluble grit" to provide them with the calcium they need for making egg-shells, and the owner has to provide this in the form of oyster shell, bought quite cheaply from a miller or pet shop. Leave a container of this available for the birds to help themselves, as it is a very individual need and should not be thought of as food.

A beginner should always buy young birds from a reputable source. Breeders advertise in local papers or *Fur and Feather*, but pedigree stock are much more expensive than utility birds. Beware of markets. Good birds can be bought sometimes but there are fearful traps for the unwary in the form of snuffly, scaly-legged, egg-bound old horrors only fit for the stockpot. Look for close-feathered, bright-eyed birds with clean legs free of any lumpiness. The beak should be dry and clean, and not cut back by a heavy-handed debeaker (de-beaking is a standard practice where poultry are to be kept in overcrowded conditions so that they peck each other). You could buy young "growers" of eight weeks or so and rear them or pay a little more and buy birds of about eighteen weeks who will come into lay three weeks later. "Point of lay" birds are expensive and seldom lay immediately because their move to new surroundings puts them off.

Some poultry-keepers like to have a male bird with their flock so that they breed their own replacement young stock (see *Eggs* for details). Others prefer to keep female birds alone, replacing them each autumn as they go into moult and stop laying. The time always comes when old birds have to be disposed of, but when you have kept poultry for some time you usually get to know other poultry-keepers or breeders who will tell you of a dealer willing to buy old stock. If you decide that you need to know how to kill a chicken, ask someone to show you. It's a knack, but it also needs some physical strength, and a bungled effort is a dreadful business. Like everything else, the expert does it best.

Hatching eggs in an incubator is dealt with under *Eggs*, but you may find that your birds take matters into their own hands by going broody. A broody hen will be found sitting in a nest box, perhaps on only one or two eggs, when all the other hens have flown up to roost at night. She will ruffle up her feathers with a 'Brrr!" noise when you approach, and try to peck you – but she won't budge off her nest. If she flies off with a lot of fuss, she is probably not quite ready to settle down to incubate her eggs and it is best to leave her for another day or two. A really motherly broody can have eggs put under her to a total of thirteen – a "natural clutch". Choose eggs not more than seven days old, from birds which have run with a cockerel, making sure they are the same size as those which the hen already has under her – or take those away and replace them with thirteen well-matched ones. The hen should cluck to them interestedly, tucking them under her feathers.

The main problem with a broody is that other hens push in beside her and add their eggs to her nest. Either mark the original thirteen eggs with crosses and take out any uncrossed eggs each evening, or "set" the broody on a prepared nest of eggs in a private run of her own such as a rabbit hutch. To do this, use plenty of clean straw, preferably lining the nest with some soft hay, hollowing it so that the eggs are securely held. Move the broody when it is getting dark, as she will be less inclined to object. Pick her up firmly and tuck her under your arm so she can't flap. Show her the eggs, and she should walk on to the nest, clucking to them, and settle down. Some hens won't be moved and it's no use battling with them. Note the date when the hen was set so that you know chicks should hatch 21 days later. A hen in a rabbit hutch has a ready -made pen for her chicks but a broody high up in a nest box or on a bale will have to be put into a run with her new-hatched brood or simply left on ground level with them to brood them as nature meant. Most hens are surprisingly capable, and very fierce in defence of their babies. All you have to do is provide chick crumbs, scattering them thinly so that the babies find them and gobbling adults don't.

Prawns(see *Sea Creatures*) **

Only pink when cooked, these are almost transparent when small, greenish when bigger. They need plenty of space and well-aerated water, but tame readily, coming to take food from the fingers. They clean up waste food obligingly.

Praying Mantis ***

If you have a taste for horror films, you will probably like these. Should by rights be spelt "preying" rather than "praying" since what appear to be clasped hands are in fact lethal spiked arms, used for catching and killing anything that moves, including other praying mantids.

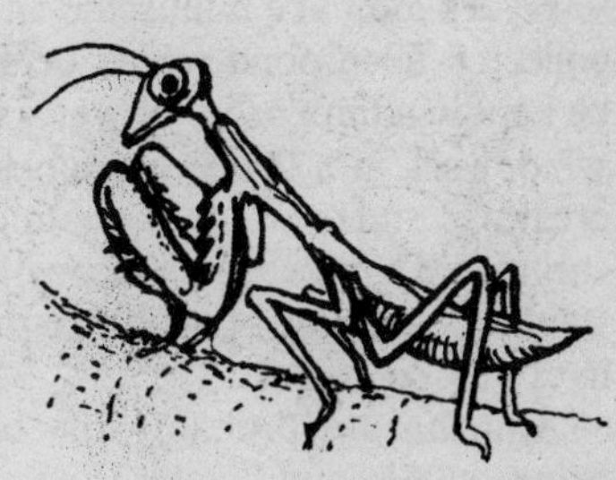

Tall sweet jars make suitable homes, and must contain twiggy branches as the egg cocoons are attached to these. The female is about 11cm (4″) long, the male much more lightly built and very vulnerable to attack by his loved one. All females must be kept alone or they will devour their companions. When contemplating mating, feed the female huge quantities of blowflies or any other living insects until she seems sated, then introduce the male. In the resulting embrace the female may well cut his head off. This doesn't matter – except to the male – as the severing of the nerve ganglion in the neck triggers the mating mechanism. In other words, the female may well be made pregnant by a technically dead husband. She will lay up to nine or ten oothecae and these cocoons should be removed to a rearing jar and put firmly on twigs. They will hatch in about five months, at least 100 nymphs emerging from each ootheca. Since they set about eating each other almost at once, this population explosion is soon controlled, and if the babies are fed plenty of fruit flies (see *Fruit Flies*) lots of them can be reared, gradually introducing small houseflies and cricket hoppers. Always provide twiggy branches to assist in skin changing and make water available in the form of soaked cotton wool.

Protozoa (see *Live Food*) *

Every puddle, pond and ditch contains millions of tiny living things which are fascinating to observe if you have or can borrow a microscope. Some, like the various species of hydra, can be seen with the naked eye. Protozoa are of vital importance as a basic foodstuff for the young of many water-dwelling species, and are quite easy to culture as an aquarium food. Never attempt to keep protozoa in tap water. The chlorine it contains is toxic to them.

Hydra look like minuscule umbrellas with no fabric and can be collected from pond weeds or submerged dead leaves. They can be kept in a tank or jar of water from their own pond, on a sunless windowsill, at a temperature below 20°C (70°F). They are greedy creatures and eat a lot of daphnia (water fleas) or cyclops or brine shrimps. They breed by depositing cysts which will survive cold weather, splitting open when the water warms up, to let young hydra emerge.

Amoeba are probably the best-known protozoa, found very commonly in pond water. Easy to culture in shallow dishes away from direct sunlight, in pond water containing three or four grains of uncooked rice.

Euglena, from farmyard pools or puddles, are commonly found near animal manure. Easy to culture in large sweet jars of pond water kept in direct sunlight. Large colonies of Euglena will soon develop in the form of a green scum. A food solution for these can be made as follows. Boil up a litre ($1\frac{1}{2}$ pints) of pond water with 30–40 grains of wheat or rice and a handful of short hay stems, adding a pinch of animal dung. Boil for ten minutes then let stand 2–3 days, stoppered. Strain and use.

Stentor is a relatively big organism, about 2mm long. It devours Euglena at the rate of 100 a minute. There are many other creatures commonly found, among them **chlamydomonas, paramecium, vorticella, colpidium, actinosphaerium** and **podophyra,** but you need help from a teacher or a specialised book in order to identify them. If your protozoa dash about too fast for observation on the microscope slide, they can be slowed down by surrounding their drop of water with a ring of 20% methyl cellulose, covering the whole thing with a coverslip.

PUPPIES (see *Dogs*) *****

PYTHONS (see *Snakes*) ***

Quails (see *Cage Birds, Poultry*) ****

The **Japanese quail** is the smallest species of domesticated poultry, its eggs being regarded as a delicacy. The even smaller **Chinese painted quail** comes in various colour forms and is apt to lay eggs all over the place without bothering to incubate them, so it is best to hatch them in an incubator or under a small broody bantam. The chicks are about the size of bumble bees. Strictly ground birds, quails are often kept in aviaries to clear up the grain and other food dropped by the perch dwellers.

Rabbits *****

There is a vast range of breeds, with widely differing weights, sizes, colours and fur types. The bigger breeds such as the **New Zealand White,** the **Californian,** the **Flemish Giant** and the **Belgian hare** are best for meat production, but as pets, smaller animals are better, being easier to lift. The **Netherland Dwarf** only weighs a kilogram (just over 2lbs) and is very docile. A little bigger, the **Dutch,** the **Himalayan** and the **English** are all attractive breeds.

It is important to buy healthy stock, bright-eyed and free from any messiness round nose or tail end. Pedigree animals suitable for show purposes can be bought from specialists advertising in the fortnightly *Fur and Feather,* the magazine of the British Rabbit Council (see *Useful Addresses*), but animals to be kept purely as pets can be bought from dealers or shops more cheaply.

In the wild, rabbits live in burrows dug with their strong claws. Kept in hutches, they have the same need for darkness and privacy, particularly when breeding, and the unused claws must be clipped from time to time to stop overgrowing. To keep their way of life as natural as possible, the ideal way to house pet rabbits is to put their hutch in an enclosure, opening the hutch door during the day so that the rabbits can run out freely to dig and exercise. The surrounding wire must be sunk into the ground to prevent escape by burrowing, and needs to be about 1 metre (3′) high, as cheerful, strong rabbits are good jumpers. The hutch itself must be off the ground to avoid damp (provide a ramp if the rabbits are let out) and must have a private nesting area with access through a pop-hole or narrow doorway. This nesting area must be supplied with plenty of straw, preferably over peat or sawdust to absorb any moisture. The living area must be well littered with peat, sawdust or cat litter.

A hutch for two rabbits should not be less than 150× 60 × 60cm (5 × 2 × 2′), and half as large again for animals of the heavy breeds. Build it of strong timber with a waterproofed felted roof, and paint the outside with polyurethane varnish or gloss paint if preferred. Tongue and grooved boarding is the ideal material as hutches must be free of draughts.

Rabbits are sociable animals and if breeding is not wanted, two or three does should be kept together. Don't try to keep bucks in a group – they are very aggressive and will fight. Rabbits will often be friendly with other species such as guinea pigs or tortoises, or even cats and dogs.

Green food fed to rabbits must not be wet. Grass or garden produce such as cabbage leaves must be picked for them the previous evening if the mornings tend to be damp or frosty and must be spread out so that they dry, not left crammed in a bucket. Hay is a valuable food and should be put in a rack so that the rabbits don't tread it about and waste it. They also eat crushed oats, flaked maize, bran and mixed corn, wholemeal bread (not white) and root vegetables. Water must always be available in a bottle drinker. Rabbit pellets can be used as a complete diet, with hay and water.

Pet rabbits must be handled frequently if they are to be tame and friendly, but painful mishandling will have the opposite effect. Never pick a rabbit up by the ears alone. A tame rabbit can be cradled safely in both arms but remember that rabbits unused to handling may kick wildly, inflicting nasty scratches. These are most safely picked up by a good handful of the loose skin on their backs or, if the owner's hand is big enough, by a firm grip across the small of the back. This holds the rabbit in a slightly head-down position, powerless to do any damage.

The doe comes into season in response to the attentions of the buck, so it is theoretically possible to mate her at any time. Never bring the buck to the doe's cage, or she will attack him. Take the doe to the buck, and watch them. If the doe sits with her bottom firmly on the floor, he cannot mate her and it is best to remove her and try again the next day. If she crouches with her tail up, he will mount her within a few minutes and, after a successful mating, will fall off sideways with a grunt. Most breeders leave the animals together for a second mating, then return the doe to her hutch.

Gestation lasts for 31 days on average and the pregnant doe should have privacy in which to prepare for her family. She will make a good nest with straw and soft hay, lining it with fur pulled from her tummy the day before the babies are born. Don't inspect the new family – you will know they are there because new-born rabbits squeak. Litters are born blind and naked, varying in number from 4–13, seven or eight being average. They are nearly always born at night. Their eyes open from 8–10 days and they will be out of the nest in a fortnight. Feed the doe as much as she can eat while she is suckling the litter and keep her living quarters clean, but don't touch the nest. Wean the young at 6–8 weeks by removing mum to a different hutch. She can be re-mated when the babies are four weeks old, thus giving her a fortnight by herself before the new litter is born.

Rabbits can breed from six weeks old but should best be mated at four months. A female stops breeding when she is three years old and the life expectancy is 6–8 years.

Any illness in rabbits is apt to be fatal, so get veterinary help at once if your pet is precious to you. Droppings should be hard, though soft pellets are taken direct from the anus and eaten, much as cattle and sheep regurgitate food to "chew the cud". Any wet messiness on a rabbit's tail end is a danger sign – remove all green food and feed on hay alone. Ear canker is caused by a small parasite and shows as a brown muckiness in the ears, causing the rabbit to scratch. Treat with canker powder or ask your vet. Myxomatosis is a man-induced disease to control wild rabbits. Its symptoms are very unpleasant, consisting of a hugely swollen head, discharging nose and eyes and obvious extreme illness. Skin haemorrhages and convulsions follow and death occurs within ten days. It is kinder to kill sufferers humanely at once. It is spread by fleas and can be brought to tame stock via hunting cats or dogs. Pet rabbits can be vaccinated against it. Sneezing and nasal discharge indicate "snuffles" often leading to pneumonia and death. Isolate any sufferer at once and consult your vet.

Always keep rabbits very clean. Fouled bedding can harbour fleas and maggots and the hutch must be cleaned out at least once a week, preferably twice. Long-haired rabbits of the **Angora** type need daily brushing and combing or their coats become a matted mess. Dont take them on unless you are prepared for a lot of work.

Make sure you can sex young rabbits with certainty, or the population will get out of hand. Sit the rabbit on your lap as shown, and gently press just above the genital opening. In a male, you can extrude the penis, whereas the female has a genital slit.

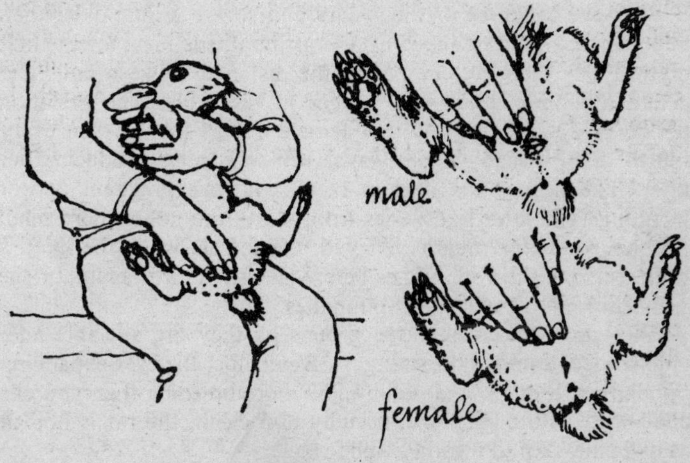

Ragworms (see *see Live Food, Sea Creatures*) **

Fast-moving, hairy marine worms, these swim and burrow. They can be dug from a muddy beach if you are quick about it, and form a splendid food for most fish. They will not foul marine tank water as lugworms do and have the intriguing habit of sharing a shell "house" with a hermit crab.

Rats ****

In many ways rats can be thought of as large mice, and treated accordingly. Obviously they must have much bigger cages, almost rabbit hutch sized, but their basic needs for exercise and hygiene are the same. Rats, however, enjoy a much wider range of foods, particularly liking household scraps of a meaty nature. Avoid cheese, as it makes them smell.

Rats live for three or four years and so are a little older than mice when they start breeding – four months is ideal for the first mating. Sex them by comparing the distance between anus and genital opening; it is much longer in the male. The gestation period is 21–26 days and the female comes into season every fourth or fifth night, and always very soon after producing her litter. The normal rat doe, in fact, is always pregnant and/or lactating. The litter size varies from 4–10, the young born blind and naked. If the male is left with the female when the litter is born, remove the young ones before she gives birth again, or she will be trying to cope with two families.

Rats can be kept in large groups as they are sociable and, unlike mice, do not fight strangers. Remember the ever-expanding population, though, and don't allow more breeding than you can cope with. Although tame, friendly and clean, the rat is not an easy animal to find a good home for.

Never be tempted to introduce wild rats to your tame stock. The native **Norwegian (brown) rat** and the less common **black rat** are undoubtedly intelligent and fascinating animals but they are apt to carry disease and are essentially uncivilised.

Don't pick rats up by the tail. They twist and struggle, often stripping the whole skin away. Grasp them firmly round the shoulders.

REPTILES

These are particularly fascinating as they are the most ancient of the four-footed animals, the living descendants of the dinosaurs. The iguana lizard, for instance, is exactly like a miniature iguanadon, and even the common tortoise has remained unchanged for millions of years.

Most reptiles lay eggs which hatch into tiny replicas of their parents, not passing through a larval stage. They are cold-blooded, which means that their bodies have no "normal" temperature as ours do, but take on the temperature of their surroundings. Thus a tortoise on a cold day is a cold tortoise, and will be sluggish and inactive. On a sunny day he will be warm and busy. For this reason reptiles hibernate throughout the winter when it is too cold for them to live actively, returning to life when it warms up. A chilly summer can be fatal to reptiles, specially young ones, as they will be too sluggish to eat properly, and will be undernourished when the autumn comes and the days grow short. To survive hibernation, an animal must have good stores of food in his body, laid down by eating well throughout the summer.

A vivarium is necessary to keep most reptiles properly (see *Vivaria* for details) and if you get their environment right, many of them will breed successfully. Living food such as worms or insects are welcomed by most species but be warned – don't use a fly spray to slow down a bluebottle so that you can catch it alive. Chemicals which kill flies will kill reptiles, too. Among other common substances which are fatal to reptiles are: paraffin, turps, DDT and other insecticides, paint solvents, iodine, sulphur and wood preservatives such as creosote. To be on the safe side, keep your vivarium free from *all* chemical substances. Dichlorvos impregnated strip (Vapona) is safe for the treatment of external parasites such as ticks and mites, but keep it out of the reptile's reach. Internal parasites can plague reptiles, but always ask your vet about this problem – the use of the wrong anthelmintic ("wormer") can be fatal.

Robins (see *Wild Birds*) **

It would be a shame (as well as illegal) to keep these nice little birds in captivity, for their willingness to share in human activity is unforced and most endearing. A cat-free, bird-encouraging household may well find a robin venturing in through the back door, and a few cheese crumbs will make him a friend for life. Robins are jealous of their territory and firmly turf out their sons and daughters to find a patch of their own, so don't expect more than a single pair to adopt you.

Salamanders (also see *Axolotl*)
These are very much like newts but stockier, with round tails rather than flat ones and five toes instead of four. Good vivarium animals (see *Vivaria*), they like each other and tend to cluster together. Less aquatic than newts, they are happy in moist places with a shelter to retire into. They are nocturnal and eat small snails, slugs, beetles, etc. The female hatches eggs inside her body, delivering live young in the form of 3cm (1″) long tadpoles, black, with a metallic gold-green sheen. These already have four legs, a long tail and three pairs of external gills. A very odd form of salamander tadpole is the **axolotl** (see under its own heading).

Salamanders were once thought to be small dragons because they secrete a white fluid in special skin glands which they can squirt to a distance of about half a metre. This is fatal to such predators as the American Bullfrog and is quite painful to humans if it should hit you in the eye, so always handle salamanders gently and at arm's length. Tame specimens won't squirt unless hurt or badly frightened.

The **fire salamander** is black, patched with brilliant orange or, in western areas of Europe, striped. It can reach 20cm (8″) long.

The **Alpine salamander** is smaller, about 10cm (4″) long, and the female produces only two live babies, so well developed as to be able to live on land.

SAND LIZARDS (see *Lizards*) *

Scorpions **

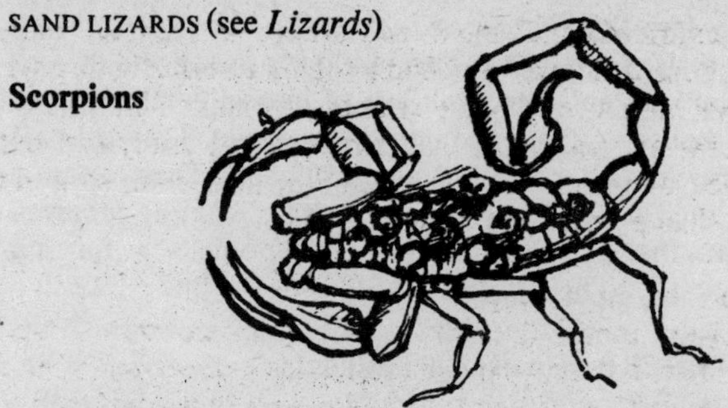

Although definitely not for pockets, these are interesting creatures to keep. They belong to the same family as spiders, the **arachnidae,** but have crab-like pincers for catching prey and for walling in their private "house" in a burrow or crevice. The sting is poisonous and although experienced scorpion-keepers pick them up by the tail, it's a dodgy business and it is much safer to use a catching box on a stick.

Scorpions need to be kept in a glass-sided vivarium with a hardboard back and narrow air gaps, warmed from underneath by an electric light bulb, giving a temperature of 21–26°C (70–80°F). They need sand to burrow in with projecting stones for shelter. Water must be provided in the form of soaked cotton wool in a open container so that the scorpions can squeeze and suck it. They only eat live food such as maggots, beetles, moths, earwigs etc. Mealworms or crickets are the easiest species to breed, ensuring a constant supply. (see *Live Food*)

In a ritual "dance" the male cleans a ground area carefully and there deposits his sperm sac ("spermatophore"). He persuades the female across the area so that she takes up the sac in her genital parts, thus fertilising her eggs. After a gestation period of 7–9 weeks, live young are born very slowly, taking about 36 hours. Each baby is vigorously brushed by the mother who then lifts it on to her back, where she carries them for 10–14 days. Then they creep off and change their skins, after which they are active little scorpions about 1cm ($\frac{1}{2}''$) long. Mum feeds them carefully with slightly-munched insects until they can catch their own. She will attack any other adult scorpion around, so dad must be removed for his own safety before the babies arrive.

Sea Anemones (see *Sea Creatures*) **

These are animals, not flowers. They are fairly hardy and do well in a marine tank if fed with small bits of mussel flesh, chopped meat or hard-boiled egg dropped into the tentacles. They also eat your small prawns, fish and other tank occupants. They release a bag of undigested food after eating, and this should be removed before it fouls the water. A dead anemone smells awful and pollutes everything, so fish it out at once.

SEA CREATURES (see individual headings)

SEA CREATURES (see individual headingr) mostly **

Animals and fish from the sea shore can be kept in tanks of sea water but they are not easy. The marine aquarist should have had plenty of experience in keeping freshwater fish and it helps to live near the sea. Inlanders need not despair – sea water can be bought in large bottles from the Marine Biological Association (see *Useful Addresses*), and synthetic sea water can be made up by any pharmacist to the formula given under *Water*. It is easier and cheaper to collect your own in stout polythene bags supported in cardboard boxes, but avoid water from smelly beaches or near chemical works or any other monstrous excrescence defiling our coastline. Collect still, clear water, not storm-tossed soup.

Salt water is highly corrosive and quickly attacks the metal of ordinary aquarium tanks. Wood is a good material, as marine creatures like darkish surroundings. Use parana pine or marine (not ordinary) ply, stuck with a resin-based marine glue, with a sheet of glass as the front wall. Water-tight glazing compound must be used for sealing – putty won't do. Any rockwork in the tank must be cemented with a compound such as Aquacrete. Old sinks etc. are useful as shallow "rock pools".

In nature, sea water moves about all the time, so a marine tank must have an aerating pump. It also needs a filtering system so that old water is constantly exchanged for new, but this is not as complicated as it seems, and specialist books contain detailed instructions. (see *Useful Books*)

Many creatures are vegetarian and eat seaweed. This also gives shelter and a sense of security to marine animals, but once parted from its roots it rots and fouls the tank, so it is best to collect young specimens firmly attached to small rocks. These will flourish, and the algae on the rocks will be appreciated, too. A thin layer of sand on the tank bottom is enough. A good marine tank should be the nearest possible replica of a natural rock pool.

Don't be tempted to put big fish in your tank – they need a lot of space and will eat everything else. Stick to small specimens and watch them carefully in case anything dies, for its body will pollute the water very badly if it is not removed at once. When catching specimens on the shore, remember not to cast a shadow across the water, or everything will hide. A net must be moved slowly through the water, a hand dipped in without splashing. The really intrepid collector goes out at night, for that is when rock-pool dwellers are most active and unafraid. Carry your catch home carefully, using a shallow container for little fish, for they are happier in a centimetre of water that splashes about and so stays oxygenated than in a stifling depth of still water. Crabs etc. are best wrapped in plenty of wet weed.

The most commonly kept and successful species are detailed under individual headings but, briefly, these are among the best things to have. Shells: limpets, whelks, mussels (essential to feed everything else), flat periwinkles and barnacles. Fish: blennies, gobies, wrasses, pollack, bass, whiting, pouting. The flat fish such as flounders and plaice are a bit unpredictable about living for long. Miscellaneous: sponges, jellyfish, anemones, starfish, sea urchins, crabs, lobsters, prawns, shrimps, octopuses and ragworm. None of these should be fed too often. Twice a week is quite enough. Daphnia, strips of white fish, crushed mussel and oocasional scraped liver or sieved hard-boiled egg are all useful.

Babies eat plankton, found only in fresh natural sea water. *Never* use fishfingers with the breadcrumbs or batter on, and avoid oily fish such as herring and mackerel, for the oil will gunge up the tank water.

Remember that sea fish are wild things. It will take them some time to settle down in a tank and the owner's head or hand appearing over the surface is a terrifying thing, enough to cause a fish to leap clean out of the tank. Be circumspect, and restrain your dabbling friends.

Breeding marine creatures is difficult, simply because full-grown specimens able to breed are too big to keep in captivity. Mullet are the easiest fish to try, but the tank must have luxuriant weed and the water must be in constant movement, very well aerated. Temperature variations are hated by all marine stock, but are absolutely fatal to developing eggs.

Seals X

These are free-living marine animals and, should you find one abandoned or injured on a beach, you must not be tempted to take it home. Leave it where it is and telephone the police or the RSPCA, asking them to contact the Seal Rescue Service (see *Useful Addresses*). Meanwhile, don't move the seal and don't fuss round it. If it seems desperately hot and dry, drape it gently with wet seaweed or, on a weed-free beach, a wet towel. Never put baby seals back in the sea. They can't manage by themselves and will only die somewhere else. Rearing these beautiful creatures is a skilled business and that is why all efforts must be made to contact an expert. More rescue centres are badly needed, and interested coast-dwellers will be welcomed, but the work is too difficult for children to take on without willing and capable adult help.

Sea Urchins (see *Sea Creatures*) **

These look like prickly balls and browse on algae and food scraps. If one is persuaded to cling to the glass front of its marine tank, the mouth with its five teeth can be clearly seen, and can be directly fed with small scraps of mussel flesh, white fish or raw meat, held in tweezers. Handle sea urchins gently, for their innards are very delicate.

Sheep X

These are not pets. Although orphan lambs are commonly reared on the bottle, they can be a great nuisance if brought up without the company of their own kind. Sheep are essentially gregarious and if their "flock" instinct is turned on to one hapless human, he or she has a large, woolly, tick-infested, bleating companion for years and years. Sheep are subject to compulsory dipping regulations, need shearing every year, and are prey to a great number of diseases and parasites. Much best left to the farmer.

Shrews **

These can be kept in captivity provided you accept that they are solitary, cross and ravenous. The **common shrew** is a bit smaller than a mouse but with a long-pointed nose. The **pygmy shrew** is tiny and the **water shrew** is larger, 9–10cm long (about $3\frac{3}{4}''$), with blackish fur and fringed legs and tail. Handle him with gloves on. Captive shrews need a big aquarium tank with plenty of peat or soil, and hay for bedding. Provide a spill-proof drinking dish and, for water shrews, a tank big enough to swim in. Shrews eat insects, earthworms, mealworms, maggots etc., consuming their own weight of food every day (see *Live Food*). The frantic owner may prefer to feed his captive a diet of shredded raw meat and oatmeal.

Shrimps (see *Sea Creatures*) **

Greenish and near-transparent, these are only pink on the fishmonger's slab. Rather unsociable in the marine tank as they like to burrow in sand, but usefully tolerant of murky water and prepared to eat anything.

Silkmoths (also see *Butterflies*) ****

Huge, fabulously beautiful creatures, surprisingly easy to breed and even a potential source of income, as there is a steady demand for "set" specimens and livestock. Not to be confused with the silkworm, which is the larva of the mulberry silkmoth, a rather dowdy white affair solely bred as a commercial silk producer. The larvae of the more spectacular silkmoths also spin silken cocoons but these are not commercially viable.

Two firms (see *Useful Addresses*) supply an international market and will send silkmoth eggs and rearing instructions through the post. It is vital to be able to supply the correct food plant for the chosen species, so make sure this is available before ordering. The eggs will arrive during autumn or winter when the trees are dormant, and must be kept in the fridge (not freezer) until the buds of the food plant break in the spring. Then bring the eggs into a warm place, and the larvae will hatch. Silkmoth caterpillars are tiny at first and must not be touched with fingers. Lift them, if at all, with an artist's paint brush and never disturb them when they are sloughing their skins. Prior to this they will stop eating and lie very still. Most species moult four times, often eating part of the shed skin, then pupate, emerging as an adult moth in the autumn. These mate and lay eggs within a few days of emergence, and die shortly afterwards.

Silkmoths have names as romantic as their appearance. The

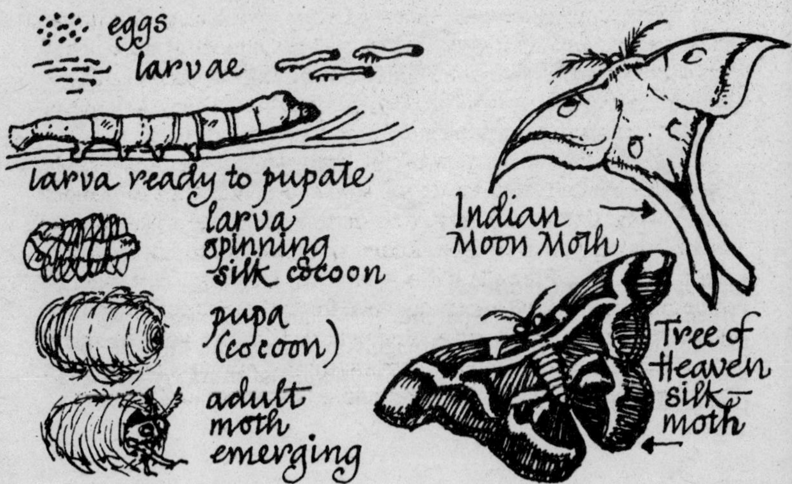

Indian Moon Moth, the **golden emperor** and the **Tree of Heaven silkmoth** are typical – and there is even a **Japanese Squeaking Silkmoth**, whose larvae chirp loudly when touched. Food requirements vary immensely but among the commonest food plants are, oak, privet, apple, hawthorn, vine, willow, lilac, cherry and virginia creeper.

Silkworms ***

These are the producers of commercial silk, domesticated thousands of years ago by the Chinese and no longer existing as a wild species. They are interesting to keep, but don't imagine you can produce a do-it-yourself silk dressing gown from a couple of hundred eggs. Silk spinning is a highly technical business – but even an amateur can produce some kind of silk.

Don't consider rearing silkworms unless you have access to a mulberry tree. The supplier will send you eggs in the autumn or winter but they must be kept in the fridge so that they don't hatch until the mulberry leaves start to break in the spring. Then they can be incubated at about 25°C (75°F) near an electric light bulb or in an airing cupboard. They hatch in about fourteen days and must not be touched with the fingers (see *Butterflies*). Transfer them to their rearing box by letting them walk on to fine strips of mulberry leaf which can be picked up. A lidded plastic sandwich box will do for the first housing, lined with paper which should be changed every day. Open boxes let the leaves dry more quickly so that fresh supplies have to be added every few hours. The larvae must never be without food as they will die almost at once. They change their skins five times before being big enough to pupate, and must never be disturbed when sloughing. They will never walk off an open tray supplied with mulberry leaves.

Silk is the substance spun by the larva when it starts to pupate (see Silkmoths illustration). It will stop eating and start to wander about when this is imminent, and must be supplied with twigs or straw among which to spin. Leave the cocoons to harden for a week then reel off the silk if it is wanted for spinning. This must be done before the adult moth emerges, for it then bites through the filament. Cocoons left untouched will develop as your breeding stock for next year. The dull, whitish moths will not live long but they lay eggs copiously, which should be kept cold until May.

SKINKS (see *Stump-tailed Skinks*) ***

Sloughing

"Sloughing" (pronounced "sluffing") is the process of shedding the old skin when a new one has grown underneath. This may seem a funny thing to do until you realise that human skin is shedding all the time. Minute flakes of skin form one of the bulkiest ingredients in household dust, believe it or not! Other mammals such as dogs and horses also renew their skin continuously, which is one reason why they need grooming if their coats are to look shiny.

Reptiles and amphibians, many insects, spiders and such invertebrates as lobsters have a different kind of skin, much tougher and more waterproof, sometimes as hard as armour. This has to be renewed from time to time, specially when the animal is still growing, so a new skin forms underneath the existing one and when it is ready the old skin comes off. In some species such as the grass snake it is shed complete, even to the transparent coverings of the eyes, and may sometimes be found by country-dwellers, looking like the empty ghost of its owner.

Many species eat their sloughed skin. It contains food elements vital to them, so they should never be prevented from doing this. Not all creatures shed their skin as a complete covering, and it does not always slip off without any effort. For this reason vivarium dwellers such as snakes and lizards should always be provided with firmly-anchored twiggy branches, for they like to rub against these and help the skin to come off.

Never be tempted to "help" a sloughing pet by pulling at his skin. It may still be firmly attached to him and you will inflict a nasty wound if you try to tear it away. Some of the snakes slough piecemeal over a long period of time and you simply have to put up with them looking tatty.

Sometimes, in abnormally dry weather, wild creatures such as frogs have difficulty in sloughing because a mild degree of dehydration makes their skins very sticky. In this case a frog will appreciate being sprinkled gently with a little water, but again, never pull at the skin.

It is quite usual for creatures about to slough to go off their food for a few days beforehand. Their colouring turns dull and "chalky" and they seem disclined to move about. Insects such as mantids will attach themselves firmly to a branch and remain motionless. They must not be disturbed, for many insect species can only slough successfully in one particular position, often depending on a branch for support.

The new skin of newly sloughed creatures is very sensitive and they should never be handled at this time. Many insects and hard-shelled water ceatures are soft and helpless after sloughing and easily fall prey to predators.

Slowworms (see *Lizards*) ***

Neither slow nor worms, these are legless lizards, about 30cm (1′) long when adult, copper-brown and shiny. Unlike snakes, they have eyelids and can blink their eyes. As reptiles, slowworms hibernate from October to March and slough their skins about four times a year. In the wild, they prefer dry terrain, heathland and hedgerows, liking to burrow under stones or in loose soil. Captive slowworms can be kept in the garden or in a vivarium containing at least 3cm (1″) of dry soil and sand, some clumps of earthy grass and rotten wood, and water in a spill-proof dish. Normal room temperature suits them during the summer months but allow them to hibernate in winter, as for snakes. (see *Hibernation*)

Diet should include insects, slugs, beetles, wireworms, woodlice etc., as well as mealworms which can be brought or bred (see *Live Food*). Slowworms mate in spring and produce live young in August or September – 6–12 tiny slowworms about 5cm (2″) long, which are at once able to catch insects. They should be removed to another vivarium as their parents may eat them. Occasionally eggs may be laid unhatched instead of hatching internally. As the young grow, males lose the black stripe they are all born with, but females sometimes retain it, and always retain a black belly. Blue spotted slowworms are males.

Slowworms are good pets, breeding readily and being very helpful in the garden. They are cheap or free both to obtain and feed, and have been known to live for 54 years in captivity! Never pick up a slowworm by its tail, or you will be left holding a wildly wriggling stump. Like the related lizards slowworms cast their tails readily.

SLUDGEWORMS (see *Live Food, Tubifex Worms*)

Slugs (see *Live Food*) *

A useful live food for many birds and reptiles, but a frightful pest in the garden, so don't culture them on Dad's lettuces. For rearing instructions see *Snails.*

Snails (see *Live Food*) **

These are rather beautiful in their way, and the young can be used as live food for birds and reptiles. Keep them in a glass jar, aquarium or vivarium, the bottom covered with a layer of gravel. This in turn is covered with moist earth and leaf mould, with dry leaves scattered on top. Rotten wood chunks provide shelter and clumps of moss serve as good bases for egg-laying. The top must be covered with glass to keep the humidity high and the container must be kept out of the sun, for this will kill snails and slugs very quickly. Feed snails on lettuce, bran and rolled oats, with a little powdered chalk to help with good shell formation.

Snakes generally ***
(Vipers X)

A lot of people are scared of snakes, so the first essential is to make sure a pet snake is securely contained. They are amazingly good escapers, adept at getting through any little hole and surprisingly able to climb out of a deep container. A glass-walled, securely-lidded vivarium is essential, suitably furnished for the species you want. (see *Vivaria*).

You can't keep poisonous snakes – they are controlled by the Dangerous Wild Animals Act. There are many species of completely harmless snakes, however, which make good pets and quickly come to know their owner. Be careful to find out what your intended pet eats before you buy it, and make sure that you will have continuous supplies of its food. Don't take on a snake on the assumption that a zoo will have it if you don't want it any more – zoos are not "dustbins" for irresponsible pet-owners, and their successful breeding policy means that they usually have enough problems in finding housing for their own stock. Find out how big your snake will be when it grows up. A baby boa constrictor may look inoffensive but he will occupy the whole of the airing cupboard later on, and leave no room for sheets. All snakes are carnivorous and all have six rows of teeth which can be shed and replaced throughout the snake's life. The flickering tongue which so many people find alarming is the snake's way of exploring his environment and shows that he is in good health. The tongue takes in minute taste droplets, conveying them to a special sensory "computer" in the roof of the mouth. This is called Jacobson's organ, and is common to many reptiles.

Glass Snakes make particularly good pets. Southern European, about 1 metre (3′) long, metallic brown with a fawn head, they are amiable, easily tamed snakes which will live happily and helpfully in the garden, eating slugs, snails and occasional mice as well as a multitude of creepy-crawlies in general. They will hibernate if left out, or can stay awake indoors if kept in a woodland vivarium (see *Vivaria*). Its name comes from the fact that the tail breaks off alarmingly easily, so catch your pet coming towards you rather than going away.

About 60cm (2′) long, **smooth snakes** are grey, brown or reddish, darkly striped on the sides of the head, and with black dots on their smooth bodies. Native to Britain and Europe, they like stony hillsides or heathland where sand lizards abound, as these form their main diet. In captivity, keep them in a desert vivarium (see *Vivaria*), and feed them on small mice etc., or whatever your particular snake has been used to in the shop or previous home. Wild specimens must not be caught, as they are now very rare and a protected species. Smooth snakes mate after hibernation, retaining the eggs in their body until hatched, so delivering live young, usually about six at a time, wrapped in thin membrane which ruptures at once. The babies are about 15cm (5–6″) long. These are easily tamed, intelligent snakes.

Female **Grass Snakes** are often over 1 metre (4′) long, the males a little smaller. They are grey, olive green or brown with patches of yellow or orange (sometimes absent in females) behind the head, and with black patches behind these. They also have rows of short vertical black bars along their sides. Reptiles, they hibernate from October to March, sloughing their skin two or three times a year (see *Hibernation, Sloughing*). They are native to Britain but most pet shop specimens come from Europe. Found mostly near water, grass snakes can also be found on heathland, commons and hillsides but they live quite happily in a garden or semi-aquatic vivarium (see *Vivaria*).

Their diet is frogs, newts, tadpoles, newt larvae, as well as worms and slugs etc. They also like fish (not sticklebacks, which have spines which stick in the throat) and sometimes mice and small birds. Frozen fish is a useful standby, but must be thawed slowly. They only eat between one and three times a week. Since grass snakes can be troubled by mites (tiny insects all over skin), it is a good idea to hang a small piece of Vapona Fly Strip in the vivarium out of snake reach, removing it when the mites have gone.

Females can breed from four years old, when they are half a metre (2′) long. They mate in the spring after hibernation and lay 10–50 eggs in the summer, choosing rotting vegetation or stable manure as a warm place where they will hatch. The eggs are tough and leathery, joined in a long string, and absorb moisture from their surroundings to swell to 3cm ($1\frac{1}{4}$″). They hatch in 6–10 weeks according to the temperature (quickest when warm). In capitivity, grass snake eggs are best put on to damp peat in a plastic box with a perforated lid, and kept in an airing cupboard as near to 27°C (80°F) as possible. If they look shrivelled, sprinkle them with water. If mould develops, take the lid off for a bit to reduce humidity. Babies are 15–20cm (6–8″) long, and slough skins before eating their first meal.

A wild grass snake will hiss and make a nasty smell when picked up, but it is harmless and will become quite tame. His vivarium must be kept well ventilated, escape-proof, securely lidded and kept out of direct sunlight, though he will enjoy an occasional bask in the sun. Provide a big dish of water that he can get right into, but not so full that he overflows it. Grass snakes can be kept awake all winter by maintaining a high temperature but are more likely to breed if allowed to hibernate.

Like all snakes, **pythons** are carnivorous. They are non-poisonous, eating their food by crushing it slowly in their jaws. Their saliva contains powerful enzymes which start the digestive process. Laboratory rats or mice are the easiest diet to obtain although pythons will eat any small mammals or frogs. A python weighing about 2kg (5 lbs) will need a small rat (about 200g) once a week.

Snakes prefer live food but rats can put up quite a fight, inflicting wounds on the snake which often go septic and always leave scars. To kill a mouse or rat, hold it down by means of a pencil across the back of the neck then lift the body vertically by the tail, thus breaking the neck. If the snake looks bored, push the corpse about a bit to stimulate his interest. Never use rodents for food that have died of poison – you will poison your python. Clean water must always be available.

Remove droppings as soon as they appear and your vivarium will stay clean. Ticks or mites can be dealt with by removing the python to a spare container, scrubbing out the vivarium and hanging a Vapona strip in it for ten days, removing it before replacing the python.

Pythons can and will vomit if they object to being handled, and they will also do this if their stomachs are not working properly – usually because their temperature is too low. Something between 26–34°C (75–95°F) is right, but high temperatures are as dangerous as low ones. Anything over 38°C (103°F) can be fatal, so avoid blazing sunshine on the vivarium.

A diet of very furry rats tends to cause constipation as snakes can't digest keratin (present in fur, hair, hoof and horn). When this happens you can feel a lump of "felt" in the snake's tummy. This can be shifted by dosing the creature with liquid paraffin through a stomach tube. Ask your vet.

Young pythons slough their skins in one piece (see *Sloughing*) but older ones slough piecemeal, which make them look a bit tatty. Don't try to "help them off with their coats", though. You may do irreparable damage.

Vipers (adders) are thick-bodied snakes, less than 60cm (2′) long. They are brownish, marked with a bold zigzag pattern in black, although their colouring varies a lot. Although they have a poisonous bite which is used to immobilise their prey, vipers are not aggressive, and will only bite if trodden on or picked up. If you are bitten, go to a doctor at once for an antidote. You won't die. Utterly unsuitable as pets, the keeping of vipers is forbidden by law and, in any case, they are prone to suicide by hunger strike if kept in capitivity. Do not kill one, though, if you meet it in the wild. They mean you no harm and are part of the natural scheme of things.

A reputable supplier (see *Useful Addresses*) will see that you choose a suitable pet snake, and there is no reason why you should not thoroughly enjoy owning a beautiful and unusual creature.

SPARROWHAWKS (see *Birds of Prey*) X

Sparrows (see *Java Sparrows, Wild Birds*) *

Wild sparrows are comon, cheeky and rather endearing, though they tend to push everyone else off the bird table and pinch all the nest boxes. Hedge sparrows are more shy and reserved than the brash Town sort, and are worth encouraging. Java sparrows are a different thing entirely.

SPAWN (see *Eggs, Frogs, Newts, Salamanders, Toads*)

Spiders **

Clean, quiet, beautiful and vastly intriguing, these are lovely creatures to study. There are thousands of species with a wide variety of different habits and appearance, so a spider found marooned in the bath should not be thrown out of the window, as it needs to live indoors.

House spiders live harmoniously (more or less) with man, building webs in his house and catching his flies, and they can be kept in closer captivity for observation. A big jar with muslin tied across the neck will make a reasonable home but most spiders will like a private matchbox into which to retire, and will need branching twigs or bits of stick as a support for web-building. Live flies are best supplied by putting pupae (see *Maggots*) in the jar, where they will hatch into nice fresh meals.

Most spiders live for two years, breeding in their second year. Mating can be tricky, as the females of some species will attack the male. Eggs are laid in a silk-wrapped cocoon in the autumn, hatching in May. The tiny spiders stay in a close cluster until able to spin their first web, and the surviving ones hide away through the winter to emerge in April, when they spin busily and eat voraciously, laying eggs in the autumn, then usually dying.

There are several interesting species of spider. The **wolf spider** spins no web but hunts prey, carrying babies with her. The **crab**

spider darts sideways, as its name suggests. The young spin long threads to carry them away in the wind, extending the colony to new territory. The **harvestman** seems to be all legs, with a tiny light body. Webless, this is most active in autumn. The **jumping spider** leaps on prey or anything which takes its fancy. **Water spiders** lay their eggs in a silk bell attched to under-water plants. This is kept filled with air by the mother spider charging her furry body with air on the surface and bringing a bubble down to the bell. These eat mosquito larvae and water insects. (also see *Tarantulas*)

Sponges (see *Sea Creatures*) **

Filter feeders, these animals are not happy in a nice clean aquarium tank where there is nothing to filter. They starve to death, in fact. Best to keep them a few days for observation only unless you have a cloudy tank. Beware of a normally yellow sponge turning green in blotches. It is about to die and stink.

Squirrels X

These are not pets, and any injured or orphaned squirrels must be released back to the wild once fit to live on their own. Ironically, they are also regarded as pests and may be shot by farmers, gamekeepers etc. The native **red squirrel** was the only species living in Britain until the **grey**, an imported upstart from America, was introduced as a zoo specimen. It bred alarmingly well and escaped (or was let out) rather frequently, thus founding wild colonies which have overrun the whole country in a mere fifty years or so. For a more pet-worthy cousin, see *Chipmunks*.

Stag Beetles **

Female stag beetles are large but hornless. It is only the male who sports the extravagant antlers and it seems sad that such handsome creatures live for only a few weeks, long enough to mate and lay eggs in the rotting base of an old oak tree. There they hatch into finger-sized larvae which munch away at rotten wood for five years before pupating and emerging as adult beetles. Given a good supply of rotting oak wood, larvae will live happily in a box indoors and can be clearly heard, munching. They need to dig a hole in the wood in which to pupate, so make sure to include some big lumps. Nobody knows much about stag beetle mating habits, so if you hatch your own, there's a chance to do some original observation.

STAGS (see *Deer, Turkeys*) X and ***

STALLIONS (see *Donkeys, Horses and Ponies*) *****

Starfish (see *Sea Creatures*) ***

Intersting. Starfish arms (usually five) are covered on the underside with hundreds of tiny "tube feet" with which the animals can grip strongly to move along or even prise barnacles or slipper limpets off rocks to be eaten. To feed a starfish, put a small piece of mussel flesh under the tip of an arm and watch it "screwed" in to the animal's central mouth. Be careful in collecting them – the arms drop off very easily, particularly with the **brittlestar**, which is aptly named. This is a black starfish with very long, waving arms, liking to live in groups and capable of catching such small things as prawns. However, it lives peacefully with other creatures its own size and is a good beginner's species to keep.

Starlings (see *Wild Birds*) *

Bossy birds, these are fun to watch in the garden although they bully everyone. They have a total lack of dignity and their efforts to imitate a song thrush always end in a raucous sqawk. Gardeners like them because they dig leatherjackets out of the lawn.

Steppe Lemmings (known in America as "sagebrush voles") ****

These are, in fact, voles, not the mass-suicide-prone Norwegian lemmings. Good house pets, steppe lemmings are mouse-sized with short, furry tails and round ears almost hidden in fur. The comon species is grey with a black dorsal stripe.

Steppe lemmings are colony dwellers and should not be kept on their own. They are affectionate, easy-breeding creatures and the male is helpful towards his children, even trying to rear them on his own should their mother die.

Roomy cages suitable for rats or hamsters are fine for lemmings, but these are burrowing creatures and should be provided with 10cm (4″) of peat or compost in which to dig. Nesting boxes are essential, well supplied with soft hay. Since lemmings are choosy creatures there should be one more nest box available than there are female lemmings.

Feed lemmings on grain, root and green vegetables, hard dog biscuits, bread and sunflower seeds. They must have hard things such as a chunk of wood to gnaw, or their teeth become overgrown, eventually killing them. Water must be available in a drinking bottle although lemmings need very little moisture. Like gerbils, they make hardly any urine and so are smell-free.

Lemmings breed prolifically, producing up to five litters a year, though they could in theory have as many as ten. The gestation period is 24–26 days, with 4–7 naked and blind babies in each litter. They emerge from the nest at about two weeks old and are weaned within a month, becoming fully mature at 8–10 weeks. London Zoo reports that lemmings breed best when kept in a busy place where there is a lot of activity for them to watch. They are active both day and night, sleeping intermittently, and do not hibernate. Their normal life span is 18–20 months.

Stick Insects ***

These defend themselves by pretending to be twigs, very successfully. There are several species but the common one is the **Indian Stick** which eats privet and will also consider ivy and other evergreens, and oak twigs.

Stick insects are nearly all virgins but breed quite happily without male assistance – a process called parthenogenisis. It is interesting that older reference books always include a description of the male stick insect, a little chap half the size of his wife, but more recent authors describe him as "rare", or doubt his existence at all.

The Indian Stick lives for about fifteen months and starts laying eggs when she reaches full adult size at about nine months, after her final skin change. The eggs are dropped casually at the bottom of the cage and hatch very slowly, taking 4–10 months but usually producing live young in the autumn. The nymphs change skins several times, hanging from a branch of the food plant to do so. If vertical space is inadequate, the soft, newly emerged insect may "set" in a cramped position resulting in deformity, so it is important to provide plenty of headroom. A tall sweet jar is ideal. Fresh privet twigs should be put in a small pot of water, the top bunged with cotton wool to prevent the insects from drowning in it.

The common Indian species is a placid creature but some of the other kinds have defence methods to watch out for. They can kick, prickle, make nasty smells and spray caustic fluid, so handle with care! They will also shed legs if picked up by a foot, but if this happens in the nymph stage the lost limb will grow again.

The **Queensland Titan** is a massive insect over 18cm (7″) long, with a spiny body and large spotted wings. The male, if he exists at all, is much smaller.

The **pink-winged stick** is very pretty, entirely parthenogenic, and glues her eggs on to a surface rather than just dropping them about. They hatch in as little as ten weeks, producing bright green, active nymphs.

The **Javanese Lichen Stick** is amazingly difficult to see, being spotted and mossy-looking. It likes rhododendron and warns off predators by flashing its bright orange wings, normally hidden behind their black and white borders.

The **Florida Stick** is dark with yellow and pink stripes. Its glossy body seems to be imitating a stick or rock rather than a living twig, and it stays on the ground. It can emit a spray which causes temporary blindness in man and generally seems to be one of nature's less pleasant jokes.

Macleay's Spectre is another huge Australian stick, the female weighing upwards of 20g (nearly 1 oz). Spiny and small winged, she is flightless but the smaller male can fly. Eggs are flung out by the female rather like stones from a catapult, possibly explaining the evasion capacity of the male.

Most of these species eat bramble, but it is wise to enquire of the shop or supplier as to the best diet for your chosen insect.

Sticklebacks (see *Fish, Live Food*) ***

These are pugnacious little fish, often found in streams or ponds, the male bearing two, three or four sharp spines on his back. There is also a much rarer ten-spined species. These are interesting fish to keep in a cold-water aquarium but must not be kept with other species as they are aggressive even among themselves. Although only 5cm (2″) long they need plenty of space and the tank must contain a good supply of water plants as the male uses these and any oddments he can find to build a barrel-shaped nest in the spring. When this is complete he finds a pregnant female

and nudges her into his nest where she lays her eggs. He will seek several females until he is satisfied with the number of eggs in his nest, and he then fertilises them thoroughly and stands guard over them until they hatch, fanning water through the nest with his fins. When the young hatch, in about a month, he continues to guard them, catching them in his mouth if they stray too far and returning them to the nest. The females ignore eggs, nest, babies and their frantic dad, who usually dies of exhaustion when his job is finished.

Sticklebacks have small mouths and do best on live food such as daphnia or microworms. Normally they are browny-green but the male turns orangey-red underneath in the spring.

Stoats X

These are like weasels, but more so. That is, they are bigger, heavier and even fiercer, being capable of killing chickens. In cold climates, they grow a white coat for camouflage in snow, but retain their characteristic white tail tip. This is the fur known as "ermine", which is used for Lord Mayor's robes and the like.

Stump-tailed Skinks (see *Lizards*) ***

These Australian lizards are 30cm (1′) long, covered with wrinkly scales, dark brown on the back but with yellow underparts, and brown spotted. Friendly and engaging animals they are easily tamed and sometimes breed in captivity, producing live young. They eat insects of various sorts but also enjoy banana and meat scraps.

Swallows (see *Wild Birds*) X

These summer visitors build their nests of mud pellets on the eves or beams of barns and other outbuildings. They feed on insects caught in flight, and when the young birds leave the nest for the first time they have to be able to fly at once. Sometimes, if a nest is inside an open-doored building, the newly flying young will misunderstand a window and keep dashing themselves against the glass. In this case, get a ladder and rescue them, releasing them at the door so that they do not lose their bearings. They feel lovely in the hand, but don't be tempted to hold them for too long, as birds easily suffer from shock. Toss them into the air away from you rather than put them on the ground, for their wings are much stronger than their legs.

TADPOLES (see *Axolotls, Frogs, Live Food, Newts, Salamanders, Toads*)

Tarantulas (see *Spiders*) *

There are many species of giant spider which are generally, but mistakenly, called tarantulas. Although they can bite, only a few such as the **black widow** are fatal, and these dangerous species would never be offered for sale as they are prohibited under the Dangerous Wild Animals Act.

Although they only eat living things such as moths and insects (see *Live Food*), giant spiders are very clean and placid by nature. Some of them have an irritant substance in their long fur which can cause a rash in those handling them, so it is as well to make careful enquiries from your supplier before buying one. They change their skins three times a year.

Terrapins (also see *Carolina Box Tortoise*) ***

These look like small, pretty tortoises but are basically water animals and have a flatter shell than a tortoise, a longer tail and webbed feet. The most hardy come from Europe, either the **European Pond Tortoise** which is black with radiating lines on the shell and yellow spots on the head and neck, or the **Spanish Terrapin**, which is grey-brown with yellow stripes on the neck. Commonly sold is the green, often very small, **elegant terrapin,** which has red marks behind the eyes and comes from North

America. The small ones are not "miniature" but young, and need great care. Always try to buy specimens more than 5cm (2″) long. Full-grown terrapins reach 15cm (6″) or more and can live outside in a garden pool, hibernating in thick mud if available. If the pool is mudless, provide a big box of moist soil at least 30cm (1′) all round the terrapin (see *Hibernation)*. Very small terrapins must live indoors in a bowl or semi-aquatic vivarium (see *Vivaria*) containing flat rocks arranged as shown. This provides a basking place above the water and shleter below. Keep the temperature at 26–29°C (80–85°F), if necessary by using an aquarium heater. A light bulb above the tank will be enjoyed. Small terrapins should be kept awake and warm through the winter.

Their diet in the wild is fish, tadpoles, small newts and frogs, earthworms, insects, spiders, snails and water plants. Pet terrapins will eat raw meat (chopped for little ones), raw fish, liver, watercress, lettuce, fresh fruit and white bread. To prevent calcium deficiency, bone meal or powdered cuttlefish should be rubbed inbto the meat. The water will quickly become foul if meat scraps are left to rot, so change it every day, refilling with water at the same temperature (terrapins hate cold baths). NOTE: Terrapins carry salmonella bacteria which cause food poisoning in man. Empty dirty water down the loo, not the kitchen sink, and wash your hands thoroughly after handling the terrapin or his food.

The male terrapin has a longer tail than the female. They can breed from 5–7 years old, mating and laying eggs. Remove the eggs carefully (don't turn them over) and put them on damp, loose soil or peat in a plastic bag or lidded jar or tin. Incubate them at 29°C (85°F) and keep them moist and undisturbed. They may hatch in 3–5 months.

Terrapins are moderately cheap to buy and keep but awful if not cleaned daily.

Toads (also see *Xenopus Laevis*) ***

These are similar in appearance and general behaviour to frogs but usually less aquatic. Most species have a warty skin which is poisonous to many predators although the grass snake doesn't seem to mind. Toads lay eggs (spawn) in long strings of jelly from April to the end of May. They seem more intelligent than frogs and a garden toad can learn to come when called. The usually-found British toad, is of course, called the **common toad** and is 8cm (3″) long if male, 10-12cm (4-4½″) if female.

The **natterjack toad** is worryingly rare and must not in any circumstances be taken from the wild. It can be distinguished from the common toad by the yellow stripe running down the middle of its back and, in the male, a blue throat. It is found, if at all, on shady heathland, for it likes to burrow.

For the vivarium, the **fire toad** is a better bet. Only 5cm (2″) long, he protects himself by standing up on his hind legs if threatened, thus showing his bright red tummy. Since he is a drab olive green on top, this startles the enemy somewhat. He tastes awful, anyway, so large lizards won't eat him. Much more active than most toads, he jumps about a lot and likes a deep pond in his vivarium so that he can float in it. (see *Vivaria*).

The **green toad** is Mediterranean, 6cm (2½″) long, bright green and long legged. Most toads crawl but this one hops.

The **clawed toad** is a water-dweller, to be kept in a tank rather

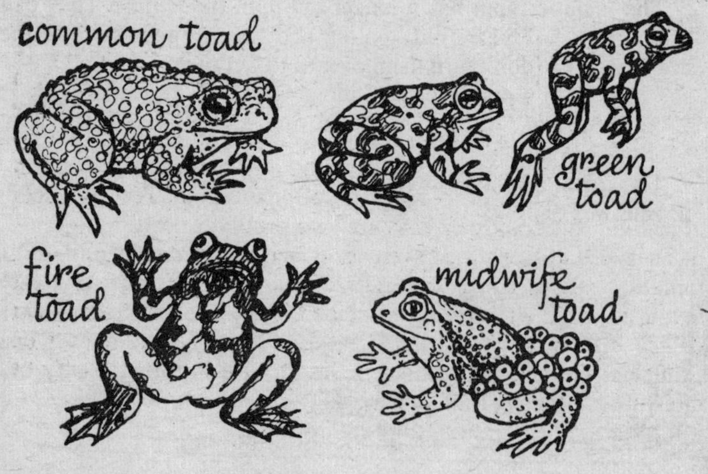

than a vivarium. He needs no artificial heat, eats worms and scraps of raw meat and has claws on his front feet. The hind ones are webbed. This rather unusual species of toad is dealt with more fully under its own heading. (see *Xenopus Laevis*)

The **midwife toad** is unlikely to show you his most interesting behaviour as, like most frogs and toads, he will not breed in captivity. A tiny animal only 4cm (1½″) long, he waits for his mate to lay her eggs then winds the long strings of spawn round his hind legs. Thus laden, he carries his embryo children about with him for three weeks, moistening them carefully in a pond from time to time, until they hatch. Although this is not normally seen in captivity, midwife toads are sometimes offered for sale complete with eggs, and these may hatch successfully in a good vivarium with a big pond. This is a European toad, pale olive in colour, spotted black and with white underparts.

Tortoises (also see *Carolina Box Tortoise*) ****

There are several kinds, but the commonly sold one is the **Mediterranean** or **spur-thighed tortoise**. The shell is formed of plates called shields, each one growing a little bigger each year to accommodate the growing animal inside. The plates are "ringed" like the cross-section of a tree, each ring probably showing a year's growth.

Long-lived reptiles, tortoises hibernate, but must be well fed during summer if they are to survive the winter (see *Hibernation*). When buying, choose a tortoise at least 18cm (5″) long (not counting head and tail). Smaller tortoises are too young to hibernate and will probably die. Always pick a heavy one, for a lightweight tortoise is thin and may have worms. A healthy specimen should look clean and active, pulling his head into his shell quickly on being touched. Always buy in the spring so as to have a whole summer's feeding before hibernation. Tortoises sold in autumn as "already hibernating" are probably dead.

Their diet is green food (lettuce, clover, grass, dandelions etc.), tomatoes, fruit, bread and milk. Squashy plums are a great favourite, often walked over to make them squashier. Some meat or fish may be liked, or tinned cat food. Water must always be available but may be ignored if the tortoise is eating plenty of moisture-containing fruit and greens.

Ideally, keep a tortoise "free range" in the garden, with a fence between him and your seedlings. If impractical, at least provide a big grassy enclosure fenced with wire or a three-brick-high wall. Watch out for digging – tortoises often stage a Colditz-type escape. A free tortoise will find himself a shelter at night, digging into matted undergrowth or long grass, but a penned tortoise must be given a house. Some owners make splendid tortoise kennels but a simple wooden box will do, provided it has a waterproof lid, a dry floor and an entrance just wide enough for the tortoise.

A male tortoise has a longer tail than a female, and a concave underside to his shell whereas hers is flat. In spring he will feel a strong mating urge and will travel miles to find a female, which is why tortoises commonly "go missing" in April or May. A pair will stay put. Round, white, leathery eggs may be laid, often in the ash of an old bonfire or in loose, dry soil. In their native Morocco the sun would hatch these, but in a farily cold country the best chance is to bury them, without turning, in a container of dry earth or sand and keep this in the airing cupboard or heated greenhouse at about 25–30°C (80–85°F). This temperature should hatch eggs in 10-12 weeks but they will take longer if it is cooler or the temperature fluctuates. Some take as long as five months – and even more never hatch at all, so don't be too disappointed if nothing happens. It is rather a triumph to breed tortoises in a chilly country. If baby torotises hatch, they will find their own way to the top of the soil, will need a drink, and in a few days will start to eat little meals of finely chopped lettuce and fruit. They can't hibernate and must be kept warm and awake indoors, in a heated vivarium. (see *Vivaria*)

Big tortoises hibernate successfully (see *Hibernation*) if well fed throughout their waking half-year. If the summer is cold a tortoise may be too sluggish to eat and must be "woken up" with some heat. An infra-red bulb about 1 metre (3′) above the ground under a perspex shelter is ideal – or just bring him indoors where it is warm, or into a heated greenhouse. He must keep eating if he is to survive the winter, but the bigger he is, the better are his chances, since he will store bigger reserves and become acclimatised to some extent to the British summer. A few raisins in his hibernation box will give him a handy breakfast should he wake before you notice him, but he will not be very active or hungry for a week or two after waking.

Tortoises are moderately cheap to buy and keep, and are the most intelligent of reptiles, friendly and responsive, quickly learning to come when called. It is a great shame that so many perish from neglect and through being exported while too young and in bad conditions. Take proper care of your tortoise and he will reward you with his surprisingly charming personality. Wash your hands thoroughly after handling him or clearing up left-over food – tortoises can harbour the salmonella bacillus, which causes food-poisoning in man.

TREE FROGS (see *Frogs*) ***

Tropical Fish (see *Fish*) *****

The setting-up of a tropical fish tank is the same as a coldwater goldfish tank, the only difference being that tropical fish live in warm water. This means that a tank heater must be incorporated, with a thermostat to make sure the water temperature stays constant. Most aquarists keep a temperature of 24°C (75°F) for their fish, but some prefer the slightly higher 25°C (78°F), claiming that this inhibits the outbreak of parasitic disease. The thermostat should be as far as possible from the heater so that it is not recording the temperature of heated water and ignoring chillier corners. A fixed thermometer is also essential as a constant visible check. More fish can be kept in a tank fitted with an aerator than one depending solely on surface oxygen, and since a tropical fish tank is "artificial" in that it depends entirely on human-provided warmth, you may as well go the whole hog and aerate it as well, preferably harnessing the aerator to serve as filter, too. Ask your aquarium shop for advice – specialist dealers know what they are talking about and are happy to help a beginner. Find out if there is a local aquarist club – members can advise you and sell you good fish specimens and second-hand equipment.

Remember that light encourages the growth of green algae, which can be removed by chemical treatment if you wish. Green water is, however, a good food for young fry and many species eat it, so don't worry about greenness unless your tank threatens to turn into pea soup. Much more worrying is the brown cloudiness which indicates pollution. If this happens you have been overfeeding your fish – a common and disastrous beginner's mistake. Give them as much as they will eat up in two minutes, no more – and leave them without food altogether for a day occasionally. If you go on holiday, feed them well with live food before hand, then forget them. They are much less likely to die of hunger than they are to be killed by an over-lavish next-door neighbour. Use any of the food mentioned in *Fish*, and see *Live Food* for notes on useful worms etc.

Most people like a "community" of mixed species in a tank, but ask your dealer to advise you on which fish can safely cohabit. Some are lethally aggressive. The usual 60 × 30 ×30cm (2 × 1 × 1′) tank can hold up to twenty small fish, and a sample selection could consist of four neons, two cardinals, two harlequins, two lemon tetra or mollies, two platies or swords, two zebras, two barbs, two guppies, one loach and one cat. These last two are bottom feeders, very useful as general cleaners-up. Water plants can be chosen freely for their beauty, as there is a wider choice available to tropical tanks than to cold-water tanks.

Most tropical fish breed readily, in a wide variety of ways, so all specimens should be kept in pairs. A weedy nursery tank is almost essential if you want to rear young fry, as adult fish eat them. The easiest breeders to start with are the live-bearers such as **mollies, guppies, swordtails** and **platies**. In this group the eggs are fertilised by the male while inside the female fish, who gives birth to live young about a month later – but the parents will eat them unless the fry are removed to another tank. The **cichlids,** on the other hand, have an elaborate courtship and nesting proce-

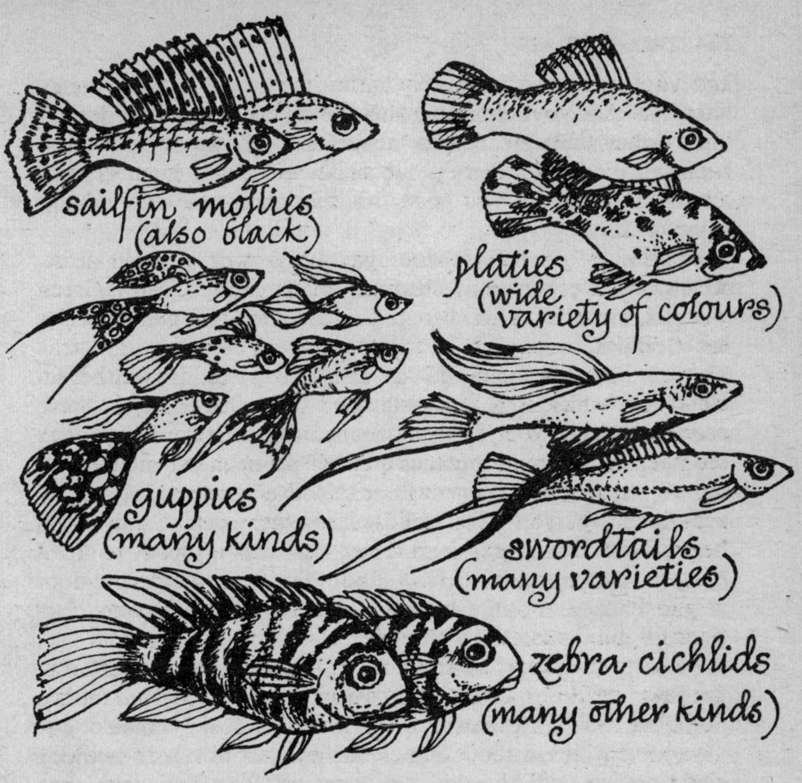

dure, fanning the eggs to provide fast-moving water and then moving the young fry daily in careful mouthfuls to a fresh nest. Several species incubate eggs in their mouths, and many use a bubble-nest. Particulary weird is the **swordtail**, whose elderly females sometimes undergo a sex change, developing into normal males able to fertilise the eggs of young female fish.

Tubifex Worms (Sludgeworms) (see *Live Food*)
2–3cm long (1–1¼″), these are red and very thin, living in mud. Ideal food for fish, they can be bought from shops or by mail order, or can be cultured by keeping them in a container of mud and water, feeding them occasionally on boiled wheat grains, boiled lettuce or bread. They should be thoroughly washed before being fed to fish or newts.

Turkeys (see *Poultry*) ***

Turkeys are the most dim-witted of domestic birds, and the white ones (**white wrolstadt**) are hopelessly unhardy. They are popular as a meat bird because they grow quickly but they are only suited to an indoor life. The male "stag" birds get so big that they would squash a turkey hen flat, so breeders use artificial insemination. Fortunately for the non-intensive back-yard poultry-keeper, it is still possible to buy **bronze turkeys** (mottled black, brown and white feathers) or, even better, the smaller, but quite hardy **Norfolk Black.**

It takes some time and patience to establish free-living turkeys. Start with an unrelated Black stag and two or three Black hens bought as young birds in their first summer. If they have been used to eating ruinously expensive turkey pellets, get them used to ordinary chicken pellets or mash, with whole wheat and plenty of fresh greenstuff. This last is particularly important for turkeys. They will eat a lot of grass and have a useful taste for nettles, but also love lettuce and fruit of all kinds. They will devastate your garden, flying over quite high fences to do so, and it may be necessary to clip one wing to prevent this. Turkeys shut in all the time are apt to eat hay or straw as a grass substitute and after a few weeks of this suicidal habit, will die of impacted gizzards. Once your birds are happy on a mixed diet, they should go through the winter without bother. If they start to sneeze or their faces swell up with sinusitis, ask your vet for help quickly, as these conditions are intractible if they persist for long.

Turkeys start to lay in the spring, young birds in May, but older ones often as early as March if the weather is warm. The eggs are white and rather pointed, brown-blotched and sometimes spotted with a kind of chalky deposit. The hens are unlikely to go broody before June and are seldom good mothers, being so stupid. However, the eggs hatch well in incubators (see *Eggs*), and the chicks are passionately fond of hard-boiled eggs, only taking to turkey starter-crumbs reluctantly. Chopped lettuce is popular from an early age. A good broody hen of a big breed such as a **Rhode Island Red** does a splendid job with turkey chicks. Parthenogenesis occasionally happens in turkeys – that is, unfertilised eggs will develop and hatch, always as male chicks.

Being a suicidal lot, young turkeys will throw themselves into tanks or buckets of water to drown, and will run about squawking dismally if it rains. Gradually the survivors gain more sense and those that attain maturity adopt quite a tough life style, roosting in trees or on walls in all weathers. Hand-reared young turkeys become convinced that their owner is their mum, and will run after you endlessly. One of mine developed a fixation about motorbikes and ran after them until restricted to the yard.

Despite their lack of brain, a family of glossy black turkeys with a handsome stag displaying like a peacock adds glamour to your poultry yard and makes the bother of rearing them worthwhile. And for furtive Christmas-dinner-fanciers, the Norfolk Black does taste absolutely delicious!

Turtles (see Terrapins) ***

There are lots of kinds, many enormous. The small, petworthy ones are more generally known as terrapins.

Vets

Just as a car owner has to allow for some repair expenses, people who keep pets must expect to need expert help occasionally if their animals are to remain healthy. In the vast majority of cases, veterinary charges average out at only a few pence a week, and yet the life or death of the animal may be at stake. So never hesitate to call your vet if your pet seems ill. Waiting to "see how he is tomorrow" may lose the vital hours when early treatment could have been started.

When you first acquire your pet, make sure you know who your local vet is, and that you have a note of his telephone number in case of emergencies. If an animal is badly injured, don't move it unless you have to, and then use a blanket held like a stretcher between two people. Never give it anything to eat or drink, and keep strangers away. Keep the animal warm but be careful how you touch it – when in great pain even the nicest of pets may snap. Above all, don't fuss.

If pills are prescribed for your pet, make sure you know how to give them – ask if necessary – and remember to give them as often as instructed, otherwise there's no point in the treatment.

Vets can do lots of things for you as well as treat sickness and injuries; routine vaccinations are needed for many pets, and some need periodic services such as claw-clipping. Vets can give you an expert opinion on an animal you are thinking of buying, advise on matters of housing and diet, and put you in touch with breeders or suppliers of livestock. They can tell you how to deal with fleas and parasitic worms, and in the sad cases of old age or hopeless infirmity, they can bring a peaceful and pain-free end. When it comes to the crunch – and it always does, sooner or later – every pet-owner's best friend is his vet.

VIPERS (ADDERS) (see *Snakes*) X

Vivaria

The vivarium is something between a cage for animals and a tank for fish, since it provides an environment suitable for reptiles or amphibians which need both earth and water. A glass aquarium tank with a tight-fitting lid is the most usual form of vivarium, and it may be one of three basic types.

1 Woodland vivarium for toads, tree frogs and salamanders. Cover the bottom of the tank with 3cm (1″) small pebbles for drainage, then 5mm ($\frac{1}{4}$″) charcoal chips to keep the soil sweet. Lastly add about 3cm (1″) or so of loamy garden soil in which you can grow some damp-loving plants such as ferns, moss, grass or tree saplings. Provide a small dish of water and some twiggy branches.

2 Semi-aquatic vivarium for grass snakes, frogs, toads, terrapins, newts and axolotls. Organise the earth as for the woodland vivarium, but put in a much larger dish of water, sunk so that it is level with the earth. Cover the bottom of the water container (which should be at least the size of an old pie dish) with sand in which two or three water plants can grow. Remember that a frog or grass snake will displace his own bulk of water when he gets in his bath, so don't fill it to the top, or the whole vivarium will get soggy when it overflows.

3 Desert vivarium for lizards, chameleons and smooth snakes. This is a dry vivarium although drinking water must be provided in a small, spill-proof dish. Cover the floor of the tank with gravel, then add a layer of sand or sandy earth. This can be planted with cacti or succulents. As in all vivaria, twiggy branches should be provided for climbing on and because these aid skin changing. (see *Sloughing*)

WATER
earth
charcoal
pebbles

1

water
earth
charcoal
pebbles
sand

2

WATER
stone
stone
sand
gravel

3

A vivarium must be kept scrupulously clean and unwanted food must be removed before it goes bad and pollutes your pet's environment. Water from a pond or stream is better than tap water as pond water is free of chlorine and contains microscopic life. The young of many species can live on this. If any animal seems ill, remove it from the vivarium at once, before it can infect the others. A 5 litre (1 gallon) jar will make a good temporary "hospital".

Many reptiles eat smaller animals – including their own young – so it is wise to see that the occupants of your vivarium are of similar size. Snakes are best kept on their own as they eat lizards and frogs. Big lizards tend to eat little ones and axolotls eat anything that moves, but some mixed groups live in peaceful harmony. Frogs, toads, salamanders and terrapins are fine together and tree frogs are quite happy with chameleons. Newts are all right with small terrapins but large terrapins are best kept alone.

If you want to catch animals such as frogs, toads and newts for your vivarium, go properly prepared, for these animals must not be carried home in hands or pockets. Take a fine-meshed net for catching, and a small enamel pie dish in which to examine your net's contents. Your captives will travel home comfortably in a jar or tin half-filled with damp moss, but remember that air holes must be punched outwards so that sharp edges do not injure the animals contained. And don't catch more animals than you need. Decide before you start how many your vivarium can house – unless it is very big, two or three are quite enough – and don't be tempted to overcrowd. Remember that the frog population is dwindling, and that each one you take is lost forever, for frogs will not breed in captivity. Responsible pet-keepers must be equally responsible about wild creatures.

Vivarium animals have delicate skins and besides, all wild creatures are badly affected by the shock of being handled, so don't touch them more than you have to. Large frogs can be grasped round the waist but most creatures are best held gently by the whole hand, letting legs and tail come between the fingers. Never pick up lizards by their tails, as tail-shedding is a common defence mechanism. Newts and other water animals are more comfortable if you wet your hand before picking them up.

Before deciding where to keep your vivarium, remember that reptiles and amphibians vary a lot in their temperature requirements. Terrapins, for instance, love a sunny windowsill but frogs would hate it and would probably die. As a rough guide, woodland dwellers like to keep cool, whereas desert creatures such as lizards need warm, dry conditions. If your animals are taken from the wild, try to match the vivarium environment as closely as you can to their natural habitat. If they are bought from a shop, notice the temperature in which the shopkeeper houses them and try to make sure that your vivarium is at the same heat. If in doubt, always ask for advice. Any reputable animal dealer will be glad to help you.

VIXENS (see *Foxes*) X

Voles **

These can be tamed with patience, except for the **water vole** which is a rat-sized, ratty-natured beast. The **Orkney vole** lives only in Orkney but the **bank vole** and its stumpier cousin the **short-tailed vole** are commonly found. Mouse-sized, they have furry ears and tails and blunt muzzles. Being determined gnawers, these need to be kept in glass or plastic cages, but need wood to gnaw. Provide a nesting box filled with hay and litter the rest of the cage with peat or sawdust. Supply water in a drinking bottle and feed your vole on oats and wheat, raw root vegetables, bits of apple and soft fruit and greenstuff in the form of grass, clover, dandelion etc. Voles also like meat sometimes, or live food such as earwigs or beetles. (see *Live Food*)

Water

All animals need water at some time and in some form, be it dew, a damp atmosphere or merely the stored water present in leaves or the flesh of other animals. Domesticated animals need water available at all times, and it must be clean. Many a cat will jump into the kitchen sink to look for clean water rather than drink from the dusty bowl on the floor.

Tap water is safe for humans and the larger animals to drink but it contains chemicals – notably chlorine – which are fatal to small organisms. It is also totally lacking in any food value and is useless for such creatures as tadpoles, young fish and newts.

Pond water, on the other hand, is teeming with microscopic life (see *Protozoa*). Any green water such as the stuff in a vase of dying flowers, a rainwater tank or a farmyard puddle, is rich in chlorophyll-bearing small animals, and these are a fascinating study if you have or can borrow a microscope. Such water is also a splendid source of food for larger animals (see *Live Food*). Never dilute green water with tap water, and never use tap water in preparing microscope slides, or you will kill your protozoa. Keep a container outside to collect rainwater or, as a last resort, use distilled water, which is completely harmless.

Marine creatures can only live in sea water, which is very complex. Sea "salt" comes from the decomposed bodies of marine life and, on average, is present at 35:1,000 parts by weight. It accummulates during the winter when there is no sunshine but when spring comes the sunlight acts on this substance to cause it to burst into life in the form of millions of microscopic creatures. This is "plankton", the food for all other young marine creatures, whose birth follows soon after the spring upsurge of plankton.

Substitute sea water can be made with commercial salts such as Tropic Marin, Eden Brine or Meersaltz. A pharmacist can make it up according to the Marine Biological Association's formula as follows:

Sodium chloride	23.447g
Magnesium chloride	4.981g
Sodium sulphate	3.917g
Calcium chloride	1.102g
Potassium chloride	0.664g

Sodium bicarbonate	0.192g
Potassium bromide	0.096g
Boric acid	0.026g
Strontium chloride	0.024g
Sodium fluoride	0.003g
Water	1 litre

This should have 10% natural sea water added to it for trace elements, and must be well aerated before use.

WATER FLEAS (see *Daphnia, Live Food*)

Weasels *

Fiercely carnivorous, the weasel is surprisingly small, a bare hand-span from nose to hind legs, with a tail half as long again. Red-brown with white underparts, he runs with a snaky, humped appearance. He eats mice, voles, beetles, frogs, small birds and even things much bigger than himself such as rats and rabbits. He can perform a strange "dance", writhing like a snake to attract curious and edible onlookers.

The Henry Doubleday Research Association claim that the weasel acts as a marvellous deterrent to moles and rabbits if kept caged in the garden, and will send plans for weasel traps and cages. They further claim that a captive weasel becomes very tame and can be a charming pet. See *Useful Addresses* if you fancy the idea, but do wear thick gloves.

Weaver Birds (see *Cage Birds*) ****

A fascinating group of birds to keep, because they construct intricate round nests, weaving them out of long grasses. Always busy and chattering, they are a sociable lot and yet are not very easy to breed. It is best to keep a male with three or four females, housing them separately from birds of other species. They need an outdoor aviary with a warm shelter for the night, and their enclosure must obviously contain growing plants and grass. They eat canary seed and millet but also need green food and plenty of insects (see *Live Food*), particularly while breeding.

Whelks or **Dog Whelks** (see *Sea Creatures*) **
Carnivorous shellfish, these eat barnacles. They are all right in marine tanks if fed with mussel flesh.

White Worms (see *Earthworms, Live Food*) *
Good food for newts, salamanders and fish, these are found in compost. Keep them in the same way as earthworms and feed them with milk-soaked bread pushed down skewer holes in soil.

WILD ANIMALS(see individual headings)
The term "wild" is a loose one, covering anything from tigers to dormice but basically meaning animals which were not bred in captivity. Reckless capturing of wild animals for pets or as trophies has reduced the number of many species to near-extinction and laws now exist to control the trade in free-living animals. A seriously interested person, however, can still find great pleasure in watching wild creatures living their own lives or in keeping them for a short time to release later. Make sure before you take any wild creature – even a newt or beetle – away from its natural home that you are going to provide it with an environment where it can continue to live happily. Be honest – if you have thought no further ahead than a jamjar or biscuit tin put the creature back where you found it.

WILD BIRDS **
For people who can't keep domestic pets (perhaps because the Council won't allow it, for instance) wild birds go a long way towards providing companionship and interest. Even a windowsill can be made attractive to birds, who will soon come to feed there, and a garden, no matter how small, can be a constant source of interest.

Cats, of course, are the great enemies of birds and any tables or drinking bowls provided should be cat-proofed as far as possible. A bird table with a single stalk is difficult for pussy to climb, specially if it is made of smooth metal. If you actually keep a cat, forget the birds. The two just don't mix.

A good bird garden should be thoroughly weedy. Tall plants such as thistles, michaelmas daisies and sunfllowers all provide tasty seed heads, and the berries of hawthorn, holly, rowan, cotoneaster and elder are all appreciated. Bird tables are best sited near overhanging trees so that birds can hop to them easily from a sheltered place rather than have to venture out to the middle of a lawn. A water supply is, of course, essential, for bathing as well as drinking, but it is best to put a brick or two in such deep containers as old sinks so that birds can easily get out as well as in.

Various species of birds have different food requirements and in winter all birds will need more fat and protein. Slugs and snails are a great favourite with thrushes – who need a stone "anvil" on which to crack shells – and bluetits will spend hours upside down pecking at a half coconut shell suspended on a piece of string. When such shells are empty they can be filled with a mixture of fat and chopped nuts. Mealworms are a splendid food (see *Live Food*) but maggots can be too thick-skinned for young birds. Bread is useful and so, of course, are the proprietary wild bird-seed mixtures.

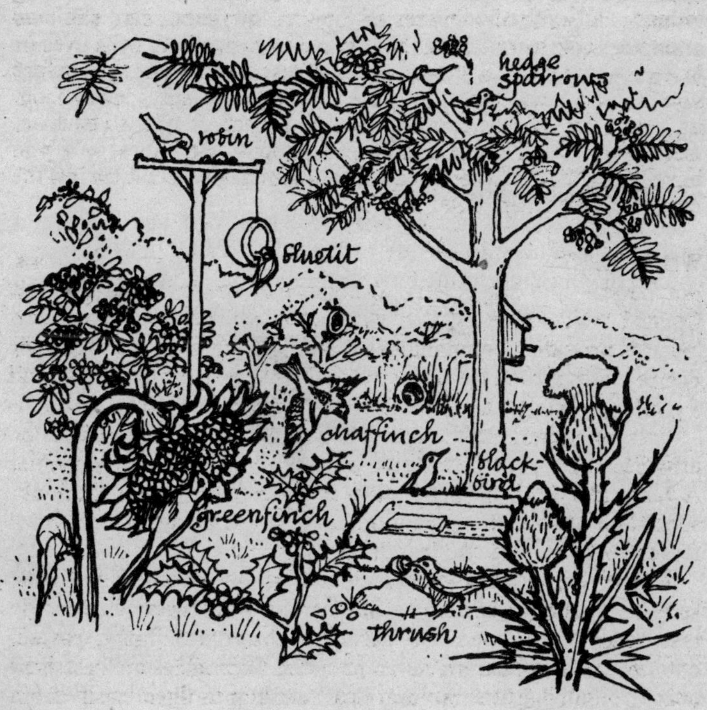

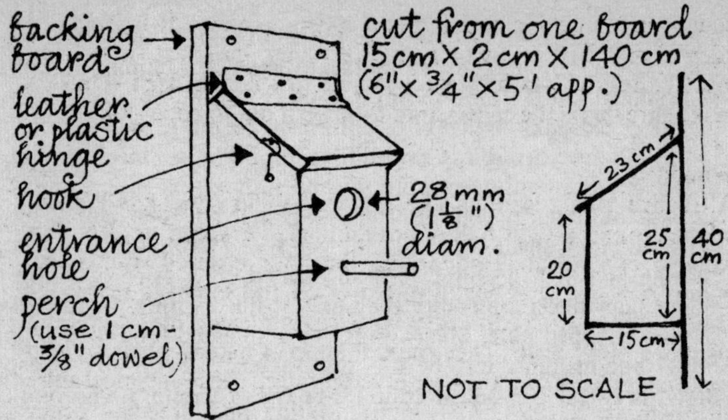

Wild birds will breed readily in a rough garden with unclipped hedges, and will often make use of hollow containers such as drainpipes, old cans on their sides or pushed into hedges. If you find a bird's nest, never try to improve its surroundings by cutting back branches to make access easier. Birds navigate by visual landmarks, rather like a driver knowing he turns left by the Rose & Crown, and if you chop off a branch it's like the pub being demolished. Confusion results.

Many species will nest in a purpose-built nest box fixed on a tree, but these must be put up early in the year (about February) to give the birds plenty of time to think about it. A nest box can be built quite easily (see diagram) or can be supplied by the RSPB (see *Useful Addresses*). It is illegal to take the eggs of any wild birds, incidentally, and baby birds found fluttering on the ground are best left to their parents as well-meant human interference usually ends in the death of the youngster. Nestlings known to be parentless need specialised care (see *Orphans*) and it is important to be certain what species they belong to before trying to feed them.

People often want to mark a bird they have reared or rescued so that they can recognise it again. The only way to do this is to fix a proper ring supplied for the species in question and the British Trust for Ornithology (see *Useful Addresses*) can advise on this. *Never* try to mark a bird by twisting an odd bit of fuse wire round its leg; almost inevitably such a bracelet will get caught in something and inflict a hideous injury. It is illegal to keep a wild bird in captivity and a young bird reared by hand must be released as soon as it is able to fend for itself – but do make sure this is the case before pushing it out to cope alone.

Woodlice **

Attractive little creatures like tiny armadillos, these roll themselves up into a tight ball when alarmed. They can be kept in a wooden box or small aquarium, the floor covered with moist, humus-rich soil, with bits of rotten wood and some big stones for shelter. Woodlice feed mostly on decaying leaves etc. in soil but also love raw potato. They need a lump of natural chalk to help make exoskeleton, for they are invertebrates. The best temperature for them is 20–25°C (70–75°F). Mating and breeding occur readily, the female carrying eggs in a pouch under her chest until they hatch in about three weeks, releasing 12–300 young.

Wrasses (see *Sea Creatures*) ***

There are several sorts of these, but they all stay near rocks, eating a variety of prawns, barnacles, crabs and smaller fish. They have the comical habit of leaning on things, particularly when asleep at night – a time when most other marine life is awake and busy. They can be seen sleeping in various odd positions, including resting on their tails with noses pointing upwards, lying flat on their sides or leaning drunkenly against a rock.

Xenopus laevis (African Clawed Toads) ***

These are interesting because they can be induced to breed at any time of the year by injecting the hormone chorionic gonadotrophin, which makes them extremely valuable laboratory animals. They are completely aquatic, living in water all the time – but, in the wild, if the pool they live in dries up, they bury themselves in the mud as though hibernating, and remain dormant until the rainy season, when the pool fills up again.

These toads can jump out of the water, so their tank must be covered. They are very slippery, with a smooth skin covered with marbled markings which become darker or lighter to match the background where the toad happens to be. They have no tongues, so use both hands to push food into their mouths.

They eat earthworms (see *Live Food)* but will also take pea-sized pieces of liver or sheep's heart. They should be fed twice a week and their tank water changed completely within 24 hours after feeding.

Xenopus toads can live for fifteen years and become sexually mature at two years old. The female is four times the size of the male and when mating they link together for several hours or even days. This is called amplexus, and ensures that each egg is fertilised as it is laid. The eggs and tadpoles should be reared in a separate tank, as the parents may eat them. No heater is needed, as toads and tadpoles are happy at about 22°C (72°F) and will tolerate a range of several degrees higher or lower, but sudden changes of temperature when changing toads from one tank to another must be avoided.

Zebra Finches (see *Cage Birds*) ****

The easiest and most domesticated cage bird, second only to canaries and budgies, these little finches are beautifully coloured with many variations on the theme of grey, cream and scarlet. They like a variety of millet seeds and some small canary seed, with plenty of green food. They need some soft food when rearing their young, which happens all the year round unless you take out the nest box to give the overworked female a rest. Aviary owners say that it is quite safe to let zebra finches out to fly freely in the late summer months; they will come back to the aviary for food.

ZONURE (ARMOURED LIZARD) (see *Lizards)* ***

Zoos

A zoo is much more than a museum with living exhibits. As increasing numbers of commercial interests persist in destroying the natural world and everything which lives in it, the people who are concerned with preserving wildlife are fighting an increasingly desperate battle to keep rare species in existence. Zoos are very much involved in this struggle to prevent extinction. The needs of their animals are studied in great detail so that an acceptable environment can be provided, and the proof of their success is that many creatures are now breeding in captivity which were previously thought impossible to breed.

Because animals are so expensive to feed and care for, all zoos are desperately short of money. Lots of people would like to do something to help to protect wildlife, and yet few of them realise that one of the most practical contributions you can make is simply to have a day at the zoo. Your entry fee is very little, considering the fascinating day it entitles you to, and it is so well used.

Although zoos depend on their visitors to provide the necessary funds to keep going, some visitors are amazingly stupid. With depressing regularity, animals have tins and bottles thrown at them. Dangerous rubbish is pushed into their enclosures and screams of laughter greet the sight of, say, an emu eating a bottle cap. Animals die as a result of the public's activities, and often have to be operated on to remove sharp objects from their stomachs. If you are at a zoo and see someone doing something stupid and dangerous, ask them not to, or tell a keeper.

The London Zoo, among others, has an excellent education department which produces useful leaflets on the care of pet animals. If you own an unusual pet (not a cat or dog) which develops an illness unfamiliar to your vet, the zoo provides a consultation service run by its veterinary experts. This is only available to vets, and not to members of the general public.

Finally, do not expect a zoo to take over a pet which you can't cope with. Space is limited, and your ex-pet is unlikely to be terribly rare – the zoo probably has two or three specimens of his species already. So be careful what you take on. A mistake can be disastrous for both owner and pet, but if you choose well – no matter how unusually – your pet can be a pleasure to you and to itself throughout its life.

Useful Books

There are, of course, many hundreds of books in addition to the few mentioned here, all of them by people who know what they are writing about. Read as much as you can about your chosen pet – all knowledge is useful.

The UFAW Handbook on the Care and Management of Laboratory Animals. Livingstone, London

A Manual of the Care and Treatment of Children's and Exotic Pets. British Small Animal Veterinary Association

Pet Animals and Society. Ed. R. S. Anderson, Ballière Tindall

First Aid and Care of Wild Birds. Ed. J. E. Cooper and J. T. Eley, David and Charles

Animals in the Home and Classroom. T. J. Jennings, Pergamon Press

The Observer Books of British Animals, Birds, Butterflies etc., Warne

RSPCA Official Pet Guides. Collins

Pets of Today series. Kaye and Ward

Tortoises and how to Keep Them. M. Knight, Brockhampton Press

Operation Tiggywinkle (about hedgehogs). Lawrence D. Hills, Henry Doubleday Research Association (see *Useful Addresses*)

Spiders, Scorpions, Centipedes and Mites. J. L. Cloudsley-Thompson, Pergamon Press

Coldwater Aquariums and Simple Outdoor Pools. Neil Wainwright, Warne

Seawater Aquaria. L. A. J. Jackman, David and Charles

Starting with Tropical Fish. Jose and John Thorne, The Aquarium Press, London

Hello to Ponies and *Hello to Riding*. Both by Jane Allen and Mary Danby, Heinemann

Pedigree Petfoods leaflets on pet care, nutrition, breeding etc. (They also supply films)

London Zoo pamphlets on the care of many species of pets

Periodicals: *Fur and Feather; Horse and Hound; Aquarist; Cage and Aviary Birds; Birds Illustrated; Exchange and Mart*

Useful Addresses

Suppliers

Register of Accredited Breeders and Recognised Suppliers, MRC Laboratory Animals Centre, Woodmansterne Road, Carshalton, Surrey

Xenopus Ltd. (Biological Suppliers), Holmesdale Nursery, Mid Street, South Nutfield, Redhill, Surrey RH1 4JY. Tel. Nutfield Ridge 2687. Crickets, mealworms, etc., reptiles, amphibians, fish and water plants.

Marine Biological Association, The Laboratory, Citadel Hill, Plymouth, Devon

T. Gerrard and Co, Gerrard House, Worthing Road, East Preston, Sussex

Philip Harris, Oldmixon, Weston-super-Mare, Avon. Also, Ludgate Hill, Birmingham 3

Biopet, 4 Windmill Road, Sunbury-on-Thames, Surrey

J. G. Animals, 19 Streatham Vale, London SW16

London Herpetological Agency, 154 Newington Butts, London SE11

Bioserve, 38–42 Station Road, Worthing, Sussex

Palmers, Parkway, Camden Towen, London NW1

The Butterfly Farm Ltd, Bilsington, Ashford, Kent TN25 7JW. Tel. Ham Street 2513. Price list 50p (deductible from first order)

Worldwide Butterflies and Lullingstone Silk Farm, Compton House, Sherborne, Dorset DT9 4QN

Welfare Organisations

RSPCA, Causeway, Horsham, Sussex. Also many local branches

RSPB, The Lodge, Sandy, Beds.

PDSA, PDSA House, South Street, Dorking, Surrey

The Blue Cross (see your local telephone book)

The Wildlife Rescue Service (specialising in birds of prey), The Old Chequers, Briston, Melton Constable, Norfolk. Tel. Melton Constable 375 (Michael Bignold)

Seal Rescue Service, 77 Gayton Road, Kings Lynn, Norfolk. Tel. Kings Lynn 4349 (George and Glenda Giles)

The Wildfowl Trust, Slimbridge, Gloucestershire GL2 7BT. Tel. Cambridge (Glos) 333 for information on many centres

British Trust for Ornithology, 2 King Edward Street, Oxford

The Hawk Trust, Loton Park, Shrewsbury, Salop. Tel. Halfway House 232

The Falconry Centre, Newent, Gloucestershire. Tel. Newent 820 286. Philip Glasier has one of the world's largest collections of birds of prey. Visitors can see birds nesting, rearing young and flying.

The Otter Trust, Earsham, nr. Bungay, Suffolk

Breed Societies and General

British Goat Society, Rougham, Bury St Edmunds, Suffolk

Donkey Breed Society, White Shutters, Exlade, Woodcote, nr. Reading, Berkshire

British Rabbit Council, Puresoy House, 7 Kirkgate, Sutton, Newark, Nottinghamshire

National Mouse Club, 6 Carlotn Gardens, Stanwix, Carlisle, Cumbria

National Rat Club, 57 Myrtledene Road, Abbey Wood, London SE2

The Kennel Club, 1 Clarges Street, Piccadilly, London W1

The Governing Council of the Cat Fancy, 28 Brendon Road, Watchet, Somerset

The Royal Pigeon Racing Association, The Reddings, nr. Cheltenham, Gloucestershire GL51 6RN

The British Beekeeping Association, 55 Chipstead Lane, Riverhead, Sevenoaks, Kent

The Henry Doubleday Research Association, Covent Lane, Bocking, Braintree, Essex

The National Foaling Bank, Mereton Stud, Newport, Salop. Tel. Newport (Salop) 2346 (Joanna Vardon)

Donkey Foaling Bank, Hillside Farm, Adlestrop, nr. Moreton-in-the-Marsh, Gloucestershire

Pedigree Petfoods, Melton Mowbray, Leicestershire LE13 1BB